FORENSIC DNA EVIDENCE ON TRIAL

SCIENCE AND UNCERTAINTY IN THE COURTROOM

FORENSIC DNA EVIDENCE ON TRIAL

SCIENCE AND UNCERTAINTY IN THE COURTROOM

Victoria Grace
Gerald Midgley
Johanna Veth
Annabel Ahuriri-Driscoll

EMERGENT™
PUBLICATIONS

The image on the cover is part of a structure made from newspapers by Peter Midgley (1921-1991).

Forensic DNA Evidence on Trial:
Science and Uncertainty in the Courtroom
Written by: Victoria Grace, Gerald Midgley, Johanna Veth and Annabel Ahuriri-Driscoll

Library of Congress Control Number: 2011921638

ISBN: 978-0-9842165-4-3

Printed in the United States of America

ACKNOWLEDGEMENTS

We are extremely appreciative of the contribution made by participants in the research discussed in this book. Our research was only made possible by their willingness to contribute their views in interviews, or to participate in discussion groups. Two reviewers made helpful suggestions for improvements to the manuscript and we extend our thanks for their expert reflections. We gratefully acknowledge the Marsden Fund of the Royal Society of New Zealand for funding this research and also for their support (contract number ESR0601).

ABOUT THE AUTHORS

Victoria Grace is Professor of Sociology, University of Canterbury, New Zealand. Her research in the field of science and technology studies includes: understandings of genetic connectedness in the context of conception through gamete donation, experiential discourses on the use of sexuopharmaceuticals, the ontology of medical visualisation, the assumptions of human genome epidemiology, and critiques of dualism in medicine and the life sciences. She also writes in the fields of social and feminist theory, and has a longstanding research interest in psychosomatics.

Gerald Midgley is Professor of Systems Thinking and Director of the Centre for Systems Studies, Business School, University of Hull, UK. He also holds adjunct appointments at the University of Canterbury, New Zealand; the University of Queensland, Australia; and Victoria University of Wellington, New Zealand. From 2003-2010, he was a Senior Science Leader at the Institute of Environmental Science and Research (New Zealand), where (amongst other things) he undertook research on public and professional understandings of forensic DNA evidence, and also investigated ethical issues associated with forensic DNA technologies. He has had over 300 papers published in international journals, edited books and practitioner magazines, and has been involved in a wide variety of technology foresight, public sector, community development and resource management research projects. He has written or edited 10 previous books including, *Systemic Intervention: Philosophy, Methodology, and Practice* (Kluwer, 2000); *Systems Thinking, Volumes I-IV* (Sage, 2003) and, with Kurt Richardson and

Wendy Gregory, *Systems Thinking and Complexity Science: Insights for Action* (ISCE Publishing, 2006).

Johanna Suze Veth is a forensic scientist at the Institute of Environmental Science and Research (ESR) in New Zealand. She is responsible for the analysis and interpretation of biological evidence recovered from crime scenes, and regularly provides expert witness testimony. Johanna is also a doctoral candidate in sociology at the University of Canterbury. Her current research interests include investigating the extent and nature of the gap between lay and professional understandings of forensic DNA technologies and identifying potential consequences for the criminal justice system.

Annabel Ahuriri-Driscoll is a Māori/public health researcher at the Institute of Environmental Science and Research (ESR) and clinical lecturer in Māori health at the Christchurch School of Medicine, University of Otago. She has been involved in research across a wide range of areas relating to Māori advancement, including traditional Māori healing, Māori-focused service, programme and organizational evaluation, and Māori community, iwi (tribe) and hapū (sub-tribe) development.

CONTENTS

FORENSIC DNA EVIDENCE ON TRIAL: SCIENCE AND UNCERTAINTY IN THE COURTROOM

A great deal of intelligence can be invested in ignorance when the need for illusion is deep.

Saul Bellow[1]

The social studies of science, or science and technology studies (STS), has systematically drawn attention to a process whereby distinct domains of technologically-embodied knowledge have been formulated, contested then socio-culturally consolidated through a process of black-boxing[2] and closure. Black-boxing typically occurs when the success of a new knowledge embodied in a technology leads paradoxically to the internal dynamics of the process being made invisible. All that is visible is what goes in and what comes out; if the result is that it is deemed to 'work', the processes involving interpretation and negotiation of its internal complexities become invisible as they are deemed matters of 'fact' (Cole, 2004; Dahl, 2010). This process has been discussed particularly in relation to the case of forensic

---

1. This quotation is cited by Michael Saks (1997-8) by way of prelude to the conclusion of his analysis of the courts' evaluation of forensic identification science.

2. The notion of 'black-box' is a metaphor taken from the fields of computing and engineering to refer to a device that has known and identifiable input and output, but the internal workings are not known or visible. The 'black-box' metaphorically refers to this zone of inaccessible and therefore non-assessable, processing. Sociologist Anthony Giddens (1997) analyses how the phenomenon whereby members of an increasingly complex society are reliant on trusting 'experts' is problematically endemic to the conditions of modernity.

DNA technologies. It is argued that this ideological process of closure, whether gradual or relatively rapid, involves a complex network of diverse elements becoming aligned in such a way as to mean not only that the knowledge in its technological manifestation is treated as a 'given', but also that the socially- and politically-invested process enabling this status of acceptance and integration is effectively rendered invisible. This phenomenon, mapped and analyzed through STS over the last couple of decades, has raised a number of vexing questions, some of which go to the heart of the field's critical endeavour. Where do social analysts of this network stand? Is there a justifiable rationale for advocating alternative constructions of knowledge as science that reject foundationalism[3], or does any such stance inevitably become an alter-foundationalism that undermines the very critique the field undertakes? These are the kinds of question that have been brought to the foreground in relation to forensic DNA evidence by Michael Lynch, Ruth McNally and Simon Cole among others[4]. We engage with these concerns and attempt to extend the debate through our research on understandings of DNA evidence in New Zealand. We analyse how the presentation of DNA evidence in the New Zealand courts is understood by criminal justice system (CJS) professionals and members of the lay public. Through this analysis, we are compelled to deconstruct the either/or of 'science' or 'common sense' and to argue for a view of 'science' that is antithetical to a social process

---

3. We use the term 'foundationalism' to refer to knowledges that rest on some form of self-evident truth, or foundation, which is not in itself considered to be derivative.

4. In particular see Lynch and McNally (1999, 2003), and Lynch and Cole (2005).

of knowledge construction that could be ideologically foreclosed.

Members of the 'lay public' are potential jurors. They also form their own judgements on the safety or otherwise of convictions or acquittals in the courts of their communities, through exposure to media of various forms. The question of how members of a jury interpret highly technical and complex scientific presentations by expert witnesses is clearly one of considerable importance, and one that exercises the legal process as efforts to broaden reliance on the presentation of such forensic evidence intensifies. Our research team[5] became increasingly aware of concerns held by various professional groups within the New Zealand CJS, about the way members of the public (and hence jurors) interpret, think about and react to the presentation of DNA evidence; particularly the perceived divergence between 'lay public' and professional understandings. After a pilot study (MacDonald, 2005) confirmed the existence of a systematically articulated concern about such a discrepancy by members of professional groups, a full research project was developed. The first objective[6] of this research was to establish the

5. Our research team comprises one university-based sociologist/ psychosocial researcher and three social science researchers within the Crown Research Institute for Environmental Science and Research (ESR), one of whom (GM) has subsequently moved to a university position in systems thinking; one team member (JV) is also a forensic scientist with ESR. ESR conducts all forensic testing in New Zealand under contract for the Police, holds the national DNA database, and is available to conduct forensic evidence testing for the defence in a case, if requested (and paid).

6. Additional objectives were to examine how different meanings and interpretations of DNA as evidence might reinforce or marginalize identifiable paradigms of justice, and to explore how forensic DNA

nature and implications of any such difference through analyzing the meanings of DNA evidence.

The expert witness for the prosecution in New Zealand formulates the presentation of DNA evidence to the court as a likelihood ratio (LR), which is a probabilistic likelihood derived from the application of a Bayesian statistical analysis (detailed further below). This presentation is different from the random match probability (RMP) used in the courts in the United States. Interpretations of the LR statement as it is read in court are the focus of our analysis of convergences and divergences of understandings of DNA evidence by members of the lay public and CJS professional groups. Firstly we situate our analysis within relevant debates identified within the social studies of science literature that pertains specifically to forensic DNA technologies; we then review the key issues in selected papers on interpretations of the LR and the RMP; the methodology of our research is outlined; and the main themes that emerged from our analysis are discussed. The points of divergence between the professional groups and members of the lay public do not fall neatly down a line between these groups. Problems with interpretation of the LR relating to the meaning of the statistics are evident across all groups. While outlining these problems we focus on the meanings of the large 'numbers' (or small probabilities) that typically characterize the LR in the talk of participants; the substitution of probabilistic terms with reductionist beliefs; the problematic consequences of associating 'science' with 'certainty'; limits

technologies are reinforcing or reshaping symbolic rituals of the CJS that in turn reflect and support dominant conceptions of justice. Māori perspectives were to be examined across these objectives. Māori are the tangata whenua or indigenous people of Aotearoa, or New Zealand.

to understanding and concerns with 'confusion'; and the ensuing implications for the use of the LR statement as currently presented in the courts. Through this analysis we interrogate the way 'science' is discursively constituted by participants across both groups, and argue that this discursive formation of knowledge is counter to what the term 'science' could most usefully mean, not only within the criminal court, but also from the point of view of critical social studies of science.

SOCIAL STUDIES OF SCIENCE, AND FORENSIC DNA TECHNOLOGIES

The very notion of 'forensic science' is a heavily contested term in the sociological literature as well as the literature dealing with forensics and the law. The concept of 'individualization' that supposedly assumes practitioners can match crime scene evidence in the form of patterns such as fingerprints, handwriting, bitemarks, voiceprints, toolmarks, firearms, tire prints, shoe prints and so on to their assumedly unique source has come under considerable scrutiny, with Saks and Faigman (2008) referring to them as 'non-science forensic sciences' (see also Cole, 2009; Kaye, 2009). DNA typing, or profiling, was introduced into the forensics arena in the mid-1980s. On the one hand legal scholars like Kaye (2009) heralded the introduction of DNA typing as the exception to the 'non-sciences' because, unlike the other forms of evidence, it has a grounding in a science (genetics), it is hypothesis testing, is replicable[7], measures specific error rates, and uses statistics to present results in probabilistic terms. On the other hand sociological researchers are more inclined to scrutinize the 'messiness' of science in any guise, ethnographically detailing its tacit knowledge, informal untidiness, and the locally-grounded specificity of its practices that belie its apparent technical and methodical

7. Foreman *et al.* (2003) note that the adequacy of match probabilities and LR values derived from single locus probe (SLP) techniques were established through large-scale experiments involving millions of between person comparisons. The SLP technique was introduced into forensic casework in 1989, followed by short tandem repeats technique (STR) in 1994 (Lambert & Evert, 1998). However, according to Foreman *et al.* (2003), STR was introduced in 1993.

purity (Locke, 2001). Through critical sociological analyses we see how the contingency of political historicity brings the 'view from nowhere' that is assumed in many of the narratives written by scientists down to its very earthy particularity. DNA typing is not exempt from this scrutiny.

A number of sociological researchers have studied how this science of DNA forensic technologies, in the UK and the US in particular, has been progressively subject to socio-political processes of closure and 'black-boxing'. With reference to Latour's actor network theory, Halfon (1998) details how technological formations are primarily social processes that can be understood as 'socio-technical ensembles' through which 'many disparate social, material and rhetorical elements are brought together in mutually constitutive ways' (p. 802). The important point in this analysis is that the very success of technologies in achieving a coherence and stability within the social is dependent on the inevitable conflicts, contradictions, and socio-political tensions that accompany their introduction being overcome. These need to be overcome, not only in the sense of achieving the appearance of being resolved, but also of being made invisible: heterogeneity, integral to the formation of technologies, is sealed over, sutured through a process of 'closure' that in the same movement 'black-boxes' and obscures the social relations that constitute its condition of possibility[8]. Halfon refers to

8. See also Lynch (1998), whose analysis of the trial of O.J. Simpson in the United States shows how, as an exceptional case, the 'dream team' defence in a sense performed the sociological work of prising open the 'black-box' of DNA profiling to expose multiple avenues for question, attack and criticism, unleashing a discourse of uncertainty that the prosecution was unable to rhetorically surpass. This 'sociology of knowledge machine' as Lynch calls it was activated at huge expense. In

social researchers such as Jasanoff (1990, 1995) who have analyzed the mediating 'boundary-work' that scientific experts for the prosecution in the legal setting perform to establish, reiterate and secure the closure he critiques. Of particular concern to Halfon is the way the DNA typing network functions to delimit the controversy of science. An effect of this process is varying degrees of distortion in the perceived unchallengeable nature of DNA evidence: lawyers deferring to scientific experts without question, and a highly selective process whereby the resources to open the black box of DNA evidence are only possible for 'spectacular' cases (such as the trial of O.J. Simpson). Through this closure and black-boxing, science and the law come to be mutually reinforcing. This is in contrast to the often-expressed view that they are incompatible or incommensurable (Halfon, 1998; Yearly, 2004), arguably because of their fundamentally different institutional framings (Luhmann, 1989).

A major contribution to the critique of black-boxing has been made by Derksen (2000, 2010), whose study focuses in depth on one dimension of DNA profiling, that of measurement error; and in particular, but not only, the measurement of the allele length[9]. Derksen

---

mundane cases the black box is, and remains, closed.

9. Specifically, Derksen (2000) states her aim as examining 'the role of measurement error in the construction of allele frequency distributions for estimating a random match probability' (p. 806). An allele is a fragment of DNA and the DNA fragment lengths at specific sites are of varying lengths between individuals. The knowledge that allele lengths are a marker of difference between individuals (except identical twins) forms the basis of DNA profiling as a forensic science. Derksen notes that the measurement of fragment length cannot be determined without error, and this is why measurement error affects calculations of frequency

details the possible sources of interpretational ambiguity in the chain of events involved in measuring allele length, and how these sources can be influential at all stages of the process, including that of characterizing a subpopulation. Accurate specification of population substructure is central to the statistical claims made to support a 'match'. Her penetrating socio-historical analysis in a sense demonstrates how an ontology of that which is measured materializes out of the measurement process. As a singular quantification of a heterogeneous process, a measurement error rate in itself functions in a similar way to the black-boxing and closure discussed by Halfon and Lynch. The number, the measurement error rate, is a reductive social construction that accomplishes its claim to objectivity through a kind of denial. It erases the subjectivity of the 'representing subject', sealing off from view the idiosyncrasies of the judgements involved in the interactions and social practices that take measure[10]. Derksen's analysis echoes Jasanoff's (1998) observation that the turbulence of intersubjective social activity is suppressed with the formal appearance of the DNA profile—a signifying gesture that covers its

---

distribution; it is also why it is crucial to determining a correspondence between a crime scene sample and a DNA sample from the suspect. This definition suggests an allele is an existing biological entity that can be measured; an alternative definition, used by forensic scientists Foreman *et al.* (2003), refers to alleles specifically as the numerical designation given to an STR profile, representing the number of times the particular core sequence of chemical components is present (p. 476).

10. In a recent commentary in *Nature*, Gilbert (2010) quotes Dan Krane, a molecular evolutionist, making a similar point regarding the possible sources of 'bias' in visualizing the peaks of the DNA profile. We place 'bias' in quotation marks to indicate it is a term used by the author and not a concept we would endorse, as it presupposes a problematic notion of uncritically accepted objectivity.

process of production. What becomes clear in the claims resulting from Derksen's work is that a highly contested field involving numerous scientific and legal players in the US during the early years of DNA profiling became progressively domesticated into a more docile consensus engineered through the formation of committees dedicated to this task. A gradual expansion of the institutional apparatus surrounding, supporting and indeed constituting DNA profiling as an area of scientific expertise for the courts was orchestrated through what Halfon would probably refer to as an actor-network, or the diverse elements of a socio-technical ensemble. As the knowledge of DNA profiling stabilized through the establishment of measurement error rates and standardized quality control procedures and protocols in laboratories, Derksen (2010) observes how new social structures emerged which in turn played their role in legitimizing the knowledge. Through this 'account of objectivity which allows it to be seen as social and natural together', Derksen (2000: 829) shows how this one very important facet of forensic DNA typing is closed by political manoeuvring at the same time that it is traversed by intense debate that is inherent to the scientific process. When the former occludes the latter, sociological investigation characteristically has played a role in attempting to de-legitimize the resulting closure.

These institutional processes of closure that have the unintended consequence of black-boxing the very workings of science to consolidate DNA profiling into a powerful social technology inevitably raise questions about how jurors, or 'fact-finders', appraise the presentation of evidence in the courts. In accordance with this sociological theorization, are they uniformly convinced by the seamless discourse of an apparently incontestable science with its awesome arsenal of molecular biology and statistics as its uncannily absent yet present support? Do they equivocate? What do they make of the 'scientific' evidence presented by an expert witness for the prosecution?

Lynch and McNally (2003) present a detailed analysis of a trial plus two appeals (one of which went to re-trial) during the 1990s in the UK, *Regina vs Adams,* to interrogate a specific legal controversy regarding the positioning of the public, as jury members, in relation to an understanding of science. This discussion points to a binary differential of 'science' and 'common sense' at play in the justice system, and provides some useful indicators of the way this binary structures the boundaries around what can be presented in court, what the expectations are of juries in terms of processes of reasoning, judgement and decision, and how 'science' is, or is not, understood. In this particular case, a young woman was raped by someone completely unknown to her, when walking home at night, in a town near London. She saw the man's face and gave some basic descriptors (approximate age, race, etc.), but this was understandably a fuzzy image given the trauma of the moment. A DNA profile taken from semen obtained through a vaginal swab was run through the DNA database but no 'match'[11] was forthcoming. The profile was stored in case an offender came to light at some future date. This is exactly what happened. Two years

11. The term 'match' is avoided by ESR witnesses in New Zealand because of its connotation of identity. The term 'correspondence' is the preferred term.

later, Denis John Adams was arrested in connection with another offence (also sexual in nature), a DNA profile was developed, and a match was made with the DNA profile from the semen from the original case. This match, the 'scientific evidence', was the only evidence in front of the jury supporting the Crown's case. There were a few other details of the case, which in fact were all somewhat supportive of the case for the defence, and were of the order of 'common sense' (this distinction between science and common sense became sedimented into the course of the trial): Adams had an alibi (this was weak as it was only able to be corroborated by his girlfriend); the complainant did not pick him out of a line-up; and Adams had a brother, so it was argued by the defence that this affected and hence compromised the statistical DNA match probability as it was presented by the expert witness for the prosecution.

During the first trial, the QC for the defence, Ronald Thwaites, attempted to overturn the presumed distinction between the scientific means of presenting and assessing the DNA evidence expressed in probabilistic terms (the figure of one in 200 million was used), and the supposedly common sense means of presenting and assessing the evidence that supported the defence case. He did this by insisting on the use of Bayes' Theorem for the assessment of all the evidence, so there was no distinction between the means of presenting, and hence assessing, the DNA evidence and the evidence supporting the defence. His point was that the 'single large statistic' already embodied within it a number of assumptions, or previous decisions, about the evidence, and that it was entirely reasonable not only to have these made visible through a

Bayesian approach, but also to apply the same method of assessment and reasoning to the other evidence.

Bayes'Theorem does not only interrogate one hypothesis and present a probability on that basis (as is the case with frequentist statistics); rather it insists on the need for two hypotheses to be tested, based on two mutually exclusive assumptions. Testing these two assumptions enables the generation of a ratio of the probability of the evidence given that the suspect is guilty versus the probability of the evidence given that the suspect is innocent. The case discussed by Lynch and McNally provides a good example of the relevance of the distinction between these two assumptions. How one assesses or evaluates the evidence that the complainant did not pick Adams out of a line-up, for example, is different depending on whether one assumes he was guilty or innocent (called the 'prior odds'): if one assumes he was guilty (his was the face she saw at the time of the assault), then the probability that she would not pick him out of a line up is going to be lower than that resulting from a different assumption—that his was not the face she saw. If one assumes he was innocent, one would assess the likelihood of obtaining this evidence differently and give it a different weighting. It was precisely this difference that Thwaites wanted the jury to consider, and to consider it *in the same terms* as the consideration given the DNA evidence. According to Thwaites's argument, this method was commensurable between the two types of evidence, but to the judge and the court it was not. After the development and implementation of a novel procedure to assist the jury to assess all the evidence in these Bayesian LR terms, they produced a guilty verdict, one that was not overturned on retrial and one further appeal that was dismissed. The important point for our discussion here is the distinction that the judge, and the subsequent Court of Appeal, drew in the process between 'science' and 'common sense', the way this distinction was presented to the jury[12] and affected the defence strategy, and what Lynch and McNally make of it.

The court (specifically the Court of Appeal) took the view that the use of a Bayesian method to arrive at a view of the evidence under different assumptions was a 'usurpation' of the jury's prerogative, indeed imperative, that they bring their own reasoned judgement, their common sense, to bear on the relationship between one element of evidence and another (see Lynch & McNally, 2003: 93). The overwhelming irony of the determination of the court, an irony observed by Lynch and McNally, that 'the apparently objective numerical figures used in the theorem may conceal the element of judgment on which it entirely depends' (cited on p. 94) is at the heart of this problematic. It is ironic because the method that Thwaites used was precisely one designed to bring the 'apparently objective' basis for arriving at the probability estimate given by the expert witness for the prosecution into the equation—to test more than one hypothesis to foreclose the possibility of voiding this vitally important subjective assumption. Bayesian statistics were not discounted by the court for their appropriateness to assisting judgements

12. To illustrate this point, we quote one of the Court of Appeal's statements cited by Lynch and McNally (2003: 94): 'More fundamentally, however, the mathematical formula, applied to each separate piece of evidence, is simply inappropriate to the jury's task. Jurors evaluate evidence and reach a conclusion not only by means of a formula, mathematical or otherwise, but by the joint application of their individual common sense and knowledge of the world to the evidence before them'.

about 'scientific evidence'. The court's determination was, however, to demarcate 'science' from 'common sense', to enclose the presentation of 'scientific evidence' within its presumed enclave of objective knowledge that is the preserve of scientific method and mathematical calculus, and to protect 'common sense' as that which is qualitatively different; an ethically robust judgement to be arrived at by members of the jury by means of a reasoning that cannot be reduced to a number that pops out, goodness knows how, from a mathematical equation or expert system. Lynch and McNally rightly interrogate this demarcation—it creates a domain of expert assertion based on a legitimated authority that is entirely constructed outside of jury deliberation. The prospect that such a domain should not be subject to challenge within the courts is a concern their paper points towards (a reiteration of closure and black-boxing); equally, if jurors have little or no technical basis for assessing evidence presented in probabilistic terms, they can be at the mercy of experts who make uncontested assertions about what the evidence shows (Lynch & McNally, 2003: 85). We take up these problematics in our discussion of the material generated by our research. Firstly, we need to outline some background to the use of Bayesian methods in the New Zealand court for the presentation of DNA profiling evidence in the form of likelihood ratios.

PROBABILISTIC EVIDENCE IN THE FORM OF LIKELIHOOD RATIOS

As Michael Saks wrote in 1997-8, 'The existence and nature of probability data are at the heart of the theory of forensic identification' (p. 15). It could also be said that how probability data are understood, justified and interpreted is at the heart of the justice delivered through forensic identification. While there has been considerable and ongoing controversy surrounding this question, Halfon (1998) makes the point that the production and interpretation of the evidence in probabilistic terms is in fact only the third of a three stage process involving DNA as evidence that could be open to contestation. The other stages that could equally be a source of controversy are firstly the collection process where samples are identified, packaged, recorded, etc., at the crime scene (a stage that came under scrutiny in the O.J. Simpson case), and secondly the specific practices in the laboratory whereby DNA samples are processed to create DNA profiles (leading to discussions about the possible presentation of 'lab error rates'). These two early stages, he argues, have been more successfully 'black-boxed' in the sense of being closed off from scrutiny[13]. In his view they are more likely, at the time of writing, to be successful sources of challenge by defence teams because the latter can insist that the protocols and procedural documentation be made available, and raise questions if this information is not forthcoming. The third stage has

---

13. Laboratories in the USA do not make their protocols available for any scrutiny (Gilbert, 2010). They are subject to audit, and this is the basis of their reporting. ESR in New Zealand does not allow perusal of their quality control protocols or any procedural documentation, but will make such material available to the defence with reference to a specific case.

been least successfully 'black-boxed', not only because the processes are more visible and presented for contestation in the courts, but, according to Halfon, largely because 'they cannot be successfully contained within a particular expert community' (p. 807). The question of how the evidence is presented to the jury, and how it is interpreted, not only by the jury but also by the expert witness, the court, the legal personnel, and those who are party to the trial remains, however, crucially important.

The supposedly incriminating profile of a DNA sample derived from a crime scene is compared to a DNA sample taken from the suspect and deemed to 'correspond', or in the more usual jargon, to 'match'. Unlike the case of fingerprint 'matches' that have been based on visual comparisons for the last 100 years, the question of whether or not a correspondence allows one to say that the two DNA samples/profiles come from one and the same source is approached with far more caution; some would say it is approached scientifically, which in our view means, amongst other things, that one cannot know in any positive and absolute sense that they come from the same source. In other words, it is not considered legitimate to rule out the possibility that a person other than the suspect was the source of the DNA and happened to have the same DNA profile as the suspect. Because of this caution, this lack of absolute certainty, the correspondence of a crime scene DNA sample with that of a potential suspect is reported statistically. The most common approaches are:

1. Random Match Probability (RMP), a calculation of the probability of obtaining the DNA evidence if it was

from an unrelated, randomly selected person from the population in question, or;

2. A Likelihood Ratio (LR), which compares this probability with the probability of obtaining the evidence if it came from a particular suspect.

The RMP is not an entirely separate statistic but comprises a part of the calculation of the LR. The random match probability (RMP) is based on the single hypothesis that the DNA sources are the same, and this is tested via the null hypothesis to arrive at a statistical probability that the evidence (the 'match') would have resulted if the suspect was *not* the source of the DNA: *given this evidence there is a one in a million chance of arriving at this 'match' if it had come from a male (or female) other than the suspect, randomly selected from the (specified) population.* This means that the null hypothesis that the two samples did not come from the same source has to be rejected. Importantly, this does not lead to the inverse statement *therefore it did come from the suspect*; the rejection of the null hypothesis is not equivalent to 'proving' support for the alternative hypothesis; there is not an inverse relation between the hypotheses. Rather, it is a matter of judgement, or opinion, as to the weight of evidence one would attribute to the 'source' hypothesis given this rejection of the null hypothesis. The same judgement is made using a Bayesian approach, which in other respects is quite different.

Those forensic statisticians who favour the application of Bayesian statistics argue that this single hypothesis, tested through frequentist statistical methods, is inadequate to the task of assessing the relationship between two

DNA profiles: a crime scene sample and that obtained from a suspect. Foreman *et al.*, for example, stated in 2003 that there is a scientific consensus that 'the weight of evidence associated with DNA findings should be reported through a LR', in other words using a Bayesian approach, and this is tracked back to Evett and Weir in 1998. The key point about a Bayesian approach is that, instead of testing one hypothesis, it considers that the relative assessment of two hypotheses is in fact necessary to do justice to the representation of evidential weight. As described in the Adams case briefly above, the odds are different depending on whether the suspect is considered *a priori* to be guilty or innocent (the prosecution and defence positions respectively). These are therefore the two hypotheses that are incorporated into a Bayesian statistical analysis of the evidence. As stated in its technical terminology by Koehler (1996): 'the likelihood ratio is the ratio of two "conditional probabilities": the probability of finding the Evidence conditioned on the truth of some Hypothesis, and the probability of finding the Evidence conditioned on the falsity of that Hypothesis' (p. 864)[14]. LRs are important in forensic science according to Bayesians because the scientist is normally not in a position to actually estimate the prior probability (Kaye & Koehler, 2003).

---

14. Bayes' Theorem: posterior odds = likelihood ratio X prior odds:

$$\frac{\Pr(Hp \mid E,I)}{\Pr(Hd \mid E,I)} = \frac{\Pr(E \mid Hp,I)}{\Pr(E \mid Hd,I)} \times \frac{\Pr(Hp \mid I)}{\Pr(Hd \mid I)}$$

where: *Hp* is the hypothesis advanced by the prosecution, *Hd* is the hypothesis advanced by the defence, *E* represents the evidence in question, and *I* represents all the other evidence relevant to the case.

Koehler (1996) is careful to distinguish between three levels of hypothesis:

1. The defendant is the source of the genetic evidence;
2. The defendant had contact with the crime scene; and;
3. The defendant is guilty of the crime (1996: 865).

Koehler insists that the latter two hypotheses are not to be deliberated by DNA analysts, as the assessment of these requires speculation about probabilities that are outside of their domain. Or in the words of Foreman *et al.* (2003: 476), 'it is now widely accepted by DNA practitioners that the consideration of prior and posterior odds in a case is the domain of the jury' (see also Kaye & Koehler, 2003). This does not mean, of course, that a Bayesian approach to assessing those odds cannot be used in the way that Thwaites attempted in the Adams trial and appeals; indeed, this is precisely the kind of method advanced through the use of the *Case Assessment and Interpretation Model* advocated by Foreman *et al.*, whereby a framework of circumstances in the specific case is assessed against a hierarchy of propositions. Thus, Bayesian logic requires the assignment of odds, which are subjectively assessed, are case-specific, and can be made explicit. Arguments for and against this approach will be taken up again in the concluding section.

Likelihood ratios are used for the assessment of the probability of the evidence for all probative DNA evidence in New Zealand. In other jurisdictions the use of LRs is often limited to the assessment of mixtures, if they are used at all. Thus, in the New Zealand court, as others, it is only the LR resulting from the first of Koehler's three levels of hypothesis, the 'source' hypothesis, which is presented by the DNA expert witness. At the time of our research, the standard presentation of expert witness evidence in New Zealand is stated as follows:

The DNA profiling results obtained from the sample indicated that the DNA could have originated from Mr. X or from another male with the same DNA profile as Mr. X at the ten DNA sites tested.

A statistical evaluation of the scientific weight of these DNA profiling results has been undertaken. I have compared the likelihood of two possible alternatives:

Either: *the DNA present in this sample originated from Mr. X,*

Or: *this DNA originated from another male, unrelated to Mr. X, selected at random from the general New Zealand population.*

Following statistical analysis it has been determined that the likelihood of obtaining these DNA profiling results is at least one million million (1×10^{12}) times greater if the DNA in this sample originated from Mr. X rather than from another male, unrelated to Mr. X selected at random from the general New Zealand population.

In my opinion the DNA evidence provides extremely strong scientific support for the proposition that the DNA detected in the sample originated from Mr. X.[15]

---

15. This final statement 'In my opinion…' is an inference drawn on the basis of the statistics presented.

This statement contains a description of the statistical evaluation of the evidence, followed by a final opinion regarding the strength of this evidence. It would appear to be vital that a jury is clear about what this statement does and does not mean, which is far from obvious based solely on this statement itself. As a ratio of the probability of the evidence given two opposing propositions (Mr. X as the source of the DNA evidence compared with a randomly selected male from the general population), the LR therefore does *not* equal the probability that the DNA originated from Mr. X. As a number of fallacious interpretations that can be made (which is also the case with the RMP method) this 'prosecutor's fallacy' sees the terms from 'the likelihood of obtaining the evidence if it came from a male randomly selected from the population' transposed to 'the likelihood of the sample coming from Mr. X given the evidence'[16]. The import of this most common fallacious interpretation is more easily understood through an example[17]. Lynch and McNally (1999) illustrate the point through comparing two questions: 'what is the probability that the animal has four legs if it's a cow?' is clearly not the same probability as its transposition 'what is the probability that the animal is a cow if it has four legs?' As Lynch and McNally point out, when dealing with DNA evidential probability statements, this transposition is more difficult to explain, identify and avoid. Kaye *et al.* (2007) refer to the prosecutor's fallacy in a study they conducted with mock jurors and the use

---

16. An early exposition and discussion of both the prosecutor's and defence lawyer's fallacy can be read in Thompson and Schumann (1987).

17. Note that Kaye *et al.* (2007: 9) claim that the prosecutor's fallacy 'is not always common', suggesting also that the defence fallacy is more common.

of RMPs, as the tendency of the jurors to confuse 'the proportion of the population that would be excluded by the DNA test with the probability that the defendant was the source of the crime-scene DNA' (p. 2). Specifically, in a case where the expert testifies that the RMP is 1%, this means there is only a 1% chance that the DNA from a randomly selected, unrelated person in the relevant general population would match the DNA profile in the crime scene sample. This was transposed by some of the mock jurors to the following: 'because there is only a 1% chance that an innocent person would match, the chance that the defendant is innocent is 1%, and hence, that there is a 99% chance that the defendant is guilty' (Kaye *et al.*, 2007: 9), or, more strictly speaking, that the defendant is the source of the DNA (Koehler's first hypothesis).

The 'defence attorney's fallacy', or 'relevance fallacy', occurs when the evidence is undervalued. For example, the juror is given the same RMP of 1% and concludes that since only 1% of the relevant population could have contributed the sample and that population is, say, 100 people, then the odds that the defendant contributed the same are only 1 in 100, so the evidence has very little value in terms of linking the defendant with the crime (Kaye *et al.*, 2007: 10). Or if the statistic was undervalued because it was interpreted to mean that one of 10 people in a population of 10 million could be the source of the DNA profile, in Thompson and Schumann's terms (1987) the evidence is undervalued because it fails to take into consideration the fact that the other 10 people are not suspects in the case.

By way of summary, Koehler (2001) reiterates that it is not legitimate to present the DNA match statistic (RMP) as a statement about:

1. The probability of guilt or innocence;

2. The source probability, and;

3. The probability that another person would match.

These points are equally relevant to the presentation of LRs. Koehler writes 'these fallacious reformulations implicitly assume knowledge about the strength of the non-genetic evidence against the matching suspect, or the size of the reference population (i.e., the number of others who could be the source)' (Koehler, 2007: 1300). This point is underlined also by Lambert and Evett (1998): there must be a clarity regarding the distinction between what the scientific evidence provides (none of 1-3 above), and the function of the court (jury), which includes assessing and deciding guilt or innocence, 'which would require knowledge of all the other evidence in the case' (p. 267).

The LR statement presented in the New Zealand court (above) concludes with a final 'opinion' that could appear to be in contravention of Koehler's point 2. But there is a crucial difference: the statement reflects the expert witness's *opinion* about the weight that he or she, with reference to his or her scientific community of peers, would be prepared to place on the suspect-as-source hypothesis being worthy of support. Although the quantitative value of the LR is pegged to the description of the level of support, it remains a judgement. Thus, the reasoning that is supportable in accordance with the evidence is to take the likelihood ratio of *obtaining the evidence* if the source was a randomly selected male unrelated to the suspect when compared to the suspect, and then make a *judgement*, acknowledged as an 'opinion', that this evidence confers strong or extremely strong support for the view that the source of the DNA is the suspect. In the words of Kaye (2009: 88), when the likelihood ratio for the hypothesis that the suspect is the source of the crime scene sample is such a very large number (in the thousands of million, or greater), 'it becomes reasonable to believe that the extensive multi-locus match indicates that the suspect is *the* source of the [in this case] stain' (see also Koehler, 1996: 870).

What does research to date tell us about how professionals within the CJS and/or jurors, or members of the lay public who might be potential jurors, interpret and appraise DNA evidence? A brief discussion of the small but growing literature on this will be followed by an introduction to our own research.

RESEARCH TO DATE

To our knowledge, no previous research has compared the interpretation and meanings of DNA evidence by those members of professional groups who have a role in the administration of the criminal justice system (lawyers, judges, forensic scientists, pathologists, police, crime scene officers, crown prosecutors), with understandings of the lay public. There has been one important study by Dahl and her colleagues (2009a, 2009b, 2010) investigating the views and discourses of lawyers in Norway regarding the presentation of forensic DNA evidence. There have been a small but not insignificant number of studies investigating the understanding of probabilistic evidence on the part of members of the lay public, often using mock jury trials. These, however, are restricted to the evaluation of RMP presentations, and therefore the present research is the first to explore interpretations of DNA evidence presented as LRs[18]. We will briefly review the most salient features of the RMP research[19].

Evidence presented in statistical or probabilistic terms is generally considered to be among the most complex and

---

18. The study by Kaye *et al.* (2007) used Bayesian statistical methods to assess the interpretations that mock jurors made of RMPs.

19. See Kaye *et al.* (2007) for a table outlining the key points of eight prior studies on RMPs (presented as either frequencies or probabilities). There is one study reported by Holmgren (2005) that used a multiplicity of methods to evaluate jury comprehension of DNA evidence, but the findings are not sufficiently well reported to warrant discussion (no indication of non-response rate to surveys, poorly stated questions, unwarranted generalization of qualitative findings, reporting on findings from focus groups and interviews with no prior presentation of these methods).

difficult (Koehler, 2001; Ivkovic & Hans, 2003). Thompson and Schumann (1987) were among the first to use a mock jury method in such a study, and they asserted that participants were susceptible to fallacious reasoning and had difficulty in detecting flaws in arguments based on statistics. This meant that evidence was frequently either overvalued or undervalued (relative to the impact prescribed by probability theory)[20].

Jonathan Koehler has conducted research on the 'psychology of numbers' involved in jurors' interpretation of DNA statistics in the US. In 2001 he claimed that little is known about how jurors think about and use DNA match statistics. He devised a series of studies to explore how different ways of presenting the statistic may affect how jurors think about the evidence. The points of difference were the way the evidence was framed—whether the identical statistic was presented as a probability (e.g. 0.1%) or as a frequency (e.g., 1 in 1,000); and the way the evidence was targeted—whether the statistic targeted a single individual (suspect) or a broader population (that is, the chance that someone else in a specific population would also have this DNA profile). He used the theory of 'exemplar cueing' to postulate that DNA RMP evidence will be more convincing if it is presented in such a way that jurors find it hard to imagine examples of others who

---

20. Kaye *et al.* (2007) note two schools of thought on the weight likely to be given to statistical presentations of evidence (RMP) in the courts: on the one hand, speculation that jurors and judges are likely to be overwhelmed by what appears to be a high degree of precision in these extremely high numbers (low probabilities) in the millions or billions and therefore afford them too much weight, and on the other hand afford them too little weight because of a greater comfort with qualitative rather than quantitative evidence.

could also match the crime scene sample, whereas they will find it less convincing if they can more easily imagine examples of others. The results showed that probability presentation was more persuasive (of accepting the source hypothesis) than frequency presentation. It also showed that single target evidence was more persuasive than multi-target, because, with single-target, it is harder to contemplate the existence of coincidental matches. Thus the combination of single-target and probability presentation reminded the jurors of the low likelihood of the source being anyone else. The potential influence of 'exemplar cueing' will be useful for consideration of our own research material below.

Kaye *et al.* (2007) also used mock jury trials as a method to examine how jurors respond to mitochondrial DNA (mtDNA) match probabilities. Although they found significant instances of both the problematic transposition (prosecutor's fallacy) and the defence fallacy, they note how jurors were 'far from overwhelmed by the mtDNA match and statistics', and observe how the mock jurors discussed the nature and extent of other evidence in the case during their deliberations (p. 37). On balance, the so-called 'fallacious' reasoning tended to favour the defence over the prosecution. Kaye *et al.* share Koehler's view that shifting the emphasis to the number of people who might also share the incriminating DNA would be beneficial to the defence; prosecutors, they suggest, might mitigate this by reiterating their emphasis on the strength of the probability regarding the evidence when the source is tested by comparison to an unrelated, randomly selected person's DNA.

METHODOLOGY

Our qualitative study required different methods to explore the issues of interest with both professional groups and with members of the lay public. We undertook a series of interviews with the former, and focus groups with the latter, ensuring as much commonality of questioning as possible between the two to enable a comparative analysis[21]. While interviews with professionals included questioning alone, the focus groups also followed the thinking of the group as forensic DNA evidence presented in the form of a mock trial unfolded on a video. This allowed us to drill down into the detail of how potential jurors responded to different elements of the presentation of forensic DNA evidence.

Interviews

Twenty-eight interviews with professionals working within or alongside the New Zealand criminal justice system were conducted between April 2008 and May 2009. The professionals were drawn from different aspects of the criminal justice system, including police, forensic scientists, forensic medical practitioners, and lawyers. The police personnel represented a variety of roles including investigative (detectives), evidence recovery (scene of crime officers) and managerial. The forensic medical practitioners included pathologists and doctors especially trained to care for victims of sexual abuse. The forensic scientists were either specialists in crime scene analysis and evidence recovery, or DNA analysis and interpretation.

---

21. Ethical approval for this research was obtained from the Human Ethics Committee at the University of Canterbury (HE 2006/132).

The lawyers practised criminal law, either for the prosecution or the defence. The length of experience of the professionals ranged from approximately a year to more than twenty years.

The research participants were recruited either through the researchers' existing contacts or by requesting a key contact within an organization to identify potential participants based on criteria related to professional role and length of service. Participation was voluntary and no incentives were offered. Each professional was interviewed at a time and location convenient to the participant, which in most instances was a private room at their place of work. Participants were provided with information about the research project in advance of the interview, and the interviewer summarized this information and answered any questions immediately prior to the interview commencing. The interviews ranged from half an hour to just under two hours, with an average length of one hour and ten minutes. The participants were interviewed using a semi-structured interview schedule (Holstein & Gubrium, 2003), which allowed for a conversational style of interaction to prompt the talk of the participants on the questions of interest (Levesque-Lopman, 2000). Meta-modelling techniques[22] encouraging specificity (Webster,

22. Meta-modeling is a term for a set of interviewing techniques that prompt the interviewee to elaborate more fully around the many unspecifieds, generalizations, things that are unsaid, etc., as these occur in narrative talk. The process postulates that speech is generated with reference to certain models, or discourses, the person uses (not consciously) as frames of reference. The questions are designed to explore concepts and their form of expression through questions such as 'What X, specifically?', 'How specifically?', 'Who's view exactly?' and 'What is this based on?'

1999) were used to further probe talk of the participants to explore the issues in depth, without leading conceptually.

The interview questions (pertaining to the focus of the current paper) covered the following areas: description of DNA typing as a technique; what it means when a correspondence is found, and how it is established; doubts or queries related to the interpretation or meaning of DNA evidence, including among their peers; interpretation of the LR statement, and views on the meanings it might have for jurors as well as other professional groups.

See Tables 1 and 2 in the Appendix for details of the interview participants.

Focus Groups

Between January and June of 2009, seven community-based focus groups were conducted to explore interpretations and understandings of members of the lay public. A pilot was conducted before these, to test our methods. The focus groups comprised two Māori groups (referred to in the discussion as M1 and M2); two European New Zealander groups (E1 and E2); two Pacific Island groups (P1 and P2); and one Chinese New Zealander group (C). We aimed to achieve a mix of participants in terms of age, gender and backgrounds to approximate the profiles of ordinary jurors.

Participants were recruited by making contact with community groups, such as those that cater to senior citizens or migrant communities, or by utilizing existing contacts in the Auckland region. In general, the participants within each group were mostly known to one another but (with two exceptions) not to the researchers. The focus groups were held at locations convenient to the group: usually a community-based location, although two of the focus groups were held at the Auckland branch of the Institute of Environmental Science and Research (where one of the researchers is based, and where much of New Zealand's forensic DNA evidence is analyzed). Refreshments were made available throughout the focus group sessions, and each participant received a $20 gift voucher to cover time and travel. Information sheets were provided in advance wherever possible. At the beginning of each group the facilitators (one or two members of the research team) restated the objectives of the research, clarified the process and obtained signed consents. The sessions ranged from an hour and twenty minutes to just under two hours, with the average length being one and three-quarter hours. The average number of participants per group was seven with actual numbers ranging between five and ten (see Table 3 in the Appendix).

The focus group sessions were split into three sections:

1. An initial discussion of DNA evidence (how people had heard about it, what they knew about DNA, etc.);

2. A presentation of a fictional (but realistic) crime scenario with a video (developed specifically for this research) of a forensic scientist giving evidence, including making use of the likelihood ratio statement, with opportunities for participants to discuss their understandings at strategic points throughout, and;

3. A final set of questions about values and principles of criminal justice, and the relationship of DNA evidence to these.

Although the method of questioning was open-ended, care was taken to ensure the discussion included consideration of the points covered in the interviews with professionals.

The second section of the focus groups bears comparison with the mock jury research used by previous researchers (Thompson & Schumann, 1987; Kaye *et al.*, 2007), but has a distinct advantage: mock juries ask participants to deliberate after presentation of all the evidence, whereas by pausing the video and allowing discussion at strategic points it was possible to generate discussion as specific terms and concepts were introduced, meaning these discursive constructions could be more closely analyzed contextually. Thus, the significance of different aspects of the presentation was made more visible than in previous mock jury research.

Transcription And Confidentiality

The interviews and the focus groups were digitally recorded and transcribed. The interview participants were then given the opportunity to review the transcription of their interviews, in order to check meaning, make changes or withdraw particular statements due to confidentiality or other reasons. Confidentiality of the lay and professional participants was maintained through the use of coded identifiers. Where interview participants made reference to specific crimes, victims, suspects or convicted criminals, or to colleagues or other professionals within the criminal justice system, we have made every effort to remove any identifying features.

Data Analysis

All material relating to the meaning of the Likelihood Ratio statement was identified from the transcripts of interviews and focus groups. A thematic analysis (Braun & Clarke, 2006) was then used to cluster the content, firstly of the interviews and secondly the focus groups. The resulting thematic clusters could be grouped under three major headings for both groups: the meaning of the LR statement; concerns about 'confusion' over interpretation; and preferred means to communicate the DNA evidence. Each of these three topic areas had numerous sub-themes that, when analyzed, revealed points of tension and contradiction as well as gaps and absences, affording questions for theoretical consideration. The qualitative data software package NVivo was used as a tool to organize and manage the data.

POSSIBILITIES AND IMPOSSIBILITIES OF MAKING SENSE OF THE CERTAIN ABSENCE OF CERTAINTY

What Does The Likelihood Ratio Statement Mean?

Focus Groups

The presentation of the LR statement in the scenario devised for the focus groups was stated in three ways, differing according to the probative value of the different samples being compared and whether the DNA profile obtained was complete or partial. The scenario involved the armed robbery of a service station by two offenders, during which a service station employee sustained minor knife wounds. The important point, however, is the extent to which responses are, or are not, contextually indexed by participants.

The first LR presentation in relation to DNA recovered from a knife blade was stated as follows: *I determined that these DNA profiling results were one thousand million times more likely if this DNA originated from Mr. Wilson* (service station employee)… with a concluding statement that *In my opinion the DNA evidence provides extremely strong scientific support for the proposition that the DNA detected in the bloodstained swab taken from the blade of the knife originated from Mr. Wilson.*

The second LR statement referred to a partial DNA profile obtained from the handle of a knife: *I determined that the likelihood of obtaining these partial DNA profiling results is at least five thousand times more likely if this DNA* *originated from Mr. Jones…* (one of two defendants) with a concluding statement that *these DNA profiling results do provide very strong scientific support for the proposition that this DNA originated from Mr. Jones.*

The third LR statement relating to a piece of chewing gum (a different crime scene sample) located at the scene was stated: *I determined that these DNA profiling results were one million million times more likely if this DNA originated from Mr. Jones…* with a concluding statement the same as the first LR statement, that is, *'in my opinion […] extremely strong scientific support…'*

There was a minority response of some focus group members to find these results 'compelling' (E1), 'pretty convincing' (M1), 'with almost 100% certainty it was Mr. Jones' (M1), 'it sounded like she [forensic scientist giving the expert testimony] was absolutely sure […] that it originated from him' (M2). When they followed up this evaluation with their reasons, it became apparent that this assessment was, in the main, supported if not formed through one of the fallacious bases for interpretation discussed earlier: 'It's a thousand million times more likely that the DNA was Mr. Wilson's' (M1) is an instance of transposition, the 'prosecutor's fallacy' and Koehler's second illegitimate interpretation (the source probability). There were numerous and repeated instances of the practice Koehler described as 'exemplar cueing' by focus group members generally (discussed below); one focus group member who was 'convinced' of the evidence found 'a chance in a billion is phenomenal… theoretically six people in the world may have similar DNA—it's highly unlikely' (E1). Focus group members did not make Koehler's third (similar) illegitimate interpretation (the probability that another person would match), in other words using quantitative terms to state a probability of another person being the source.

Among those who voiced their conviction about the strength of the evidence, there was less tendency to make Koehler's first illegitimate interpretation (the probability of guilt or innocence)—rather, the contextual features of the case were taken to be important to this assessment: 'They found the knife at the gas station, and it had the guy's DNA on it, who got stabbed. I'm not surprised really. So, one in a billion's good enough for me, convincing for me' (E1). In the following quotation this probability of guilt is uttered, but is then retracted on further consideration: 'So they were one thousand million times more likely to have committed the crime as opposed to—or had been on the scene or not—maybe they were locals and they spit their chewing gum out of the car' (M2). There was also noteworthy caution voiced by a few regarding what conclusions could be drawn from the evidence: 'it doesn't say he was there at the time, but that it originated from him' (M2). This participant went on to explicitly reject Koehler's first fallacious interpretation, and to make it clear that they as interpreters would only be able to make an assessment of guilt or innocence on the basis of combining the DNA evidence with the other evidence in the case. It was also said 'you can't just do it [assess your evidence] in an isolated manner' (E1), or 'in circumstantial cases like this [...] where one single evidence cannot nail the guy' (C).

This minority strand of conviction appeared to be buttressed by the idea that the only reason that there was this 'uncertainty', this almost but not 100% certainty about the 'match', was because the 'science' had not yet completed (or could not complete) the comprehensive research necessary to be able to then pronounce a positive identification of samples with absolute certainty: 'the only way to get perfect results is to test everybody in the world, so…' (M1); 'that DNA profile of everybody that hasn't been, haven't been done [making it] a lot harder to come out with an accurate testing' (P2); and 'they can't be certain without testing everyone in the universe' (M2). This idea was indeed presented to the focus groups in the fictional scenario evidence: 'Because we haven't tested the entire population of New Zealand, or even the world, we can't discount the possibility that two unrelated people might have the same DNA profile at the sites we look at completely by chance'[23]. The notion that if 'they'/'we' could 'test the world's population' then we would have certainty, represents a view of science that assumes absolute knowledge is possible, even if in theory, and that it is approached in an additive way. This view runs counter to the view of the scientific process we suggest is at stake here: on the contrary, science progresses through a process of on-going critique. That is, science is never able to be 'certain' about what is 'true' in the sense of obtaining absolute positive knowledge, and is only able to be 'certain' about what is 'false'. Here we do not mean 'proven false' in the tradition of scientific falsification (Popper, 1959), but rather we refer to false discourses

---

23. According to a National Academy of Sciences' 1992 recommendation, the '"scientific underpinnings of DNA typing"—[include] the fact that "[e]ach person's DNA is unique (except that of identical twins)…"' (Kaye, 2009: 90). Strictly speaking, however, even if everyone in the world was tested to produce a 'DNA profile at the sites we look at' it does not rule out the possibility that, were a 'match' found, there may be differences on those sites not 'looked at'.

as those asserting a certainty of truth. Even if the whole population of the world could be mapped for the entire human genome of each individual (not feasible according to Kaye, 2009), this mapping would still contain error rates and the ontological postulates would remain postulates. The practice of science does not presuppose an onto-epistemology that would enable a positive knowledge reflecting truth and absolute certainty.

Being persuaded by the evidence, or deliberating with articulate caution, was not the predominant response of focus group members. Their making sense practices involved much by way of questioning, speculating, proffering suggestions and then retracting them, all of which indicated a willingness to attempt to make sense of the evidence, to produce a discursive engagement with this defined field, but with the critical reflection necessary to be aware that they were far from confident in their attempts. This led to lots of laughter and jokes throughout most of the focus groups, possibly a means of deflecting, and at the same time pointing to, the implications of being so unsure (what if they really were jurors in a case?).

Whether persuaded by the evidence or not, transposing was one of the main 'logics' used to attempt to make sense of the LR. There was in fact no instance where a discussant referred to the exact wording of the LR statement to highlight that it referred to the likelihood of obtaining the evidence rather than the likelihood of the suspect being or not being the source. Nor did they query the implications of what this might mean. Transposition of the conditional probability was repeated throughout, along the lines of 'it's likely to come from Mr. Jones a million million times more than any other male member of the New Zealand public' (M1); and 'it's one million, million times more likely to be his than anyone else's in the world' (E1). This is effectively referring to the Posterior Odds Ratio, which can only be assessed, not via a transposition, but through incorporation of additional information (Prior Odds) that the DNA evidence as such cannot provide (see Koehler, 1997).

To make sense of, and try to assess the significance of this transposed interpretation meant repeated recourse to Koehler's 'exemplar cueing'; in other words, the question of how many other people could it have been was invoked frequently to be able to assess the meaning of the LR. For example: 'there's a possible 5,000 people in New Zealand who could have handled the knife as opposed to himself. And if you think that Auckland has got, say, a third of NZ's population, that means it could be another, I don't know, 1,250 people in Auckland that could have handled the knife as opposed to himself, or Wellington, that would have less' (E1); 'I mean the amount of people that we've got here' (E2); and with reference to the wording given by the expert witness, 'it would be nice to say "there are only 6,000 million people in the world here, so based on that information I've just given you […] I can surmise that no-one else in the world would have this DNA' (E1). Although a Bayesian LR depends, at least partly, on the size of the class that shares the identifying characteristics (in this case a database containing allele length comparisons of a random sample of 20,000 New Zealanders), to engage in the form of reasoning entered into in these examples is, according to Saks and Koehler (2008), another important source of fallacious interpretation. They give a number of examples of exactly this manoeuvre in relation to DNA evidence and claim that 'Although markers that rarely occur *might* be unique, it is a fallacy to infer uniqueness from profile frequencies simply because they are smaller than the number of available objects' (p. 204)[24].

Attempts were made to think about match exemplars even with the one in a billion number given: 'you might find it somewhere in the world but somewhere that's not in New Zealand [...] so you can't be looking at somebody in China, sort of saying "well, he's got the same DNA, it could be him"' (E2). Another avenue for 'exemplar cueing' involved consideration of relatives being the source of the DNA. In some of the focus groups there was an intense focus on this question of relatedness, invoking a diversion from the 'randomly selected unrelated male' to a question about the possibility of the crime-scene DNA being that of a relation to the suspect: 'it's narrowed it *down* to Mr. Jones but it doesn't actually say that it *was* Mr. Jones, because it could have been Mr. Jones's brother, or [...] sister' (P1). The view that an untested possibility of relatedness detracted from the persuasiveness of the evidence is not inconsistent with avenues that do need to be explored on a case by case basis. Where it did diverge, however, was when the assumption was made that related people should be identified and tested in preference to 'an unrelated male randomly selected from the New Zealand population'.

---

24. Koehler (1996) makes the following comment that helps clarify this point in relation to RMPs: to say 'we would expect to see this profile in approximately one out of every one-trillion people [...] is not identical to the probability that someone else exists who shares the observed profile. Although it may be extremely unlikely that a single randomly-selected person would share a DNA profile with another person, it may be quite likely that others share this profile' (p. 861).

There was one thematic strand running through and across the focus groups: an acknowledgment of an inability to understand or make sense of the LR statement. For example, 'that would have just went straight over the top of my head' (M2); 'the biggest difficulty of all is understanding this jargon' (E1); *'GM: So what does this mean to you? Response:* blah, blah, blah, blah' (P1); and 'the juror would not understand this' (C). These represent just a few responses of those who had their own reflection on their (or others') inability to make sense. There were other instances, however, where interpretations were made that were thoroughly divergent from the theoretical premises of the LR, but where the speaker appeared confident in their interpretations: 'I think it's a shame that they picked one unrelated person somewhere in New Zealand to compare the DNA with, when they could have got someone within the vicinity [...] they picked someone else in New Zealand, which is a bit far-fetched' (E1); 'because if you've got three suspects, if it doesn't match them, at least you've got a better idea than saying it could belong to Mr. Jones or another unrelated blah, blah...' (M1); and 'where do they get their numbers from if they're going to broaden it out to everybody when this case is only involving three different sorts of DNA?' (P2).

Focus group members did not tend to refer in any repeated or systematic way to possible confounding variables, such as human error, laboratory mistakes, or error margins involved in measurements. Koehler (2001) pointed to a classic literature on inferential thinking, observing that people do not reason in probabilistic but in heuristic ways. In his view this accounts at least partially for the 'mental short cuts and other rules of thumb' (p. 1299) like transposition and 'exemplar cueing' to evaluate quantitative, probabilistic evidence. Our question is—does it matter? What are the implications? We will pick up these questions in the final discussion. We turn now to the meaning of the LR to members of the professional groups interviewed.

Professionals

The professionals interviewed were a diverse group. Some had backgrounds in biological sciences and could give a detailed description of the process (or 'technique') of DNA typing. None, however, were statisticians. Koehler (1996: 876) makes the point that even 'experts do not understand likelihood ratios', noting that frequently, in his observations, they fail to appreciate that the probability that the defendant is the source of the DNA material is not a question that is answered by the DNA evidence without making unfounded assumptions about the strength of the non-DNA evidence. Elsewhere Koehler (1995) observes that the appropriate interpretation of LRs is not intuitively obvious, noting that a body of literature indicates they are easily confused with posterior probabilities. The tendency to transposition was indeed evident repeatedly amongst the professionals we interviewed. In fact, this was the predominant source of fallacious reasoning. It took the form of both asserting the source probability (Koehler's second interpretation noted above), and asserting the probability that another person would 'match' (Koehler's third). Examples of Koehler's second interpretation: 'this profile is one million, million times greater to have come from Mr. X' (l.10); 'that's a very roundabout way of saying the DNA came from Mr. X' (l.22); and 'I guess you could just say it's at least one million, million times more likely to be him than anyone else, but I don't know' (l.25). Examples of Koehler's third: 'because statistically one million, million times—basically there couldn't be another person in NZ with that DNA profile *if* they're unrelated' (l.28); 'if we take somebody else chosen at random as opposed to this fellow, the chances of it being the person chosen at random are x million' (l.29); and 'It's a possibility that it's him compared to […] a less than one millionth to a million chances it's someone else' (l.5).

Most of the instances of transposition came from members of the Police; there were also some instances from members of the prosecution. These were transpositions that were also often accompanied by statements of conviction about the 'match': 'that's a positive match' (l.12); 'we've got a DNA hit' (l.27); 'it's Mr. X's DNA' (l.28); and 'what really spins my wheels is when I […] hear the word "million" used, because then, to me I go: bingo' (l.6). The transpositions were also often accompanied by claims concerning the guilt of the suspect: 'that's why I say extremely strong scientific support that the offender is the accused' (l.27); and 'what the ESR are telling us is that Mr. X is the offender' (l.13). Here the speakers are falling into the trap of Koehler's first fallacious interpretation, with the latter attributing it to ESR.

Some statements were unclear, and others indicated considerable confusion about the statement. For example, a forensic medical practitioner said 'what it's really saying is what are the chances of that DNA coming from Mr. X […]; they've given the statistical answer to that in terms of one, at least one in one million, million' (l.14). A Police manager proposed a question that the crown prosecutor could ask of an expert witness to clarify for the jury: 'it's extremely rare for someone else to have this DNA type other than the offender, is that what you mean?' (I.23).

By comparison to the lay participants, there was less evidence of 'exemplar cueing' amongst the professional groups. This may be attributed to their different roles relative to members of the lay public, although equally this may have been an artefact of the different context for generating talk about the LR statement. With only one exception, the forensic scientists who perform the DNA typing, compile a report for the court and perform as expert witnesses, provided either an appropriate interpretive response in accordance with Bayesian theory (avoiding the fallacies), or simply repeated the 'judgement' that the evidence is indicative of extremely strong scientific support that the suspect is the source of the DNA. Apart from the forensic scientists, only two of the respondents overall (a forensic pathologist and a defence lawyer) recognized the complexity of the specificity of the LR statement and were explicit about the fact that they did not know what it meant.

Who's Confused?

Focus Groups

A striking element to the focus group discussions was repeated examples of surprised recognition that the result of a DNA profiling exercise did not produce a certainty. This surprise was reiterated over and over: 'it's not really 100%' (E2); 'I thought DNA was like pretty concrete [...] it's not as concrete as I thought' (M2); 'because there's still, you know, a reasonable amount of doubt' (M2); 'But it's still not hard evidence' (P2); 'it's not conclusive, is it?' (P1); 'it's basically saying it's not always sure, it's not 100%' (P1); 'it's not conclusive' (P2); 'I thought the DNA would be the final, the final, the final sure, yeah' (P2); 'I expected DNA to be more specific' (P2); and 'well, it's not conclusive' (C). When discussing the uncertainty, focus group members pointed to the language used by the expert witness, saying that the wording expressed doubt and ambiguity: 'they don't make it sound like they're very confident in anything with all the "coulds" and "likelihoods"; it's almost like they don't really know…' (E1); 'she's not very convinced in it herself'[25] (E1); 'the language she used, like "may" or "possible"…' (M2); 'But she's carefully worded it, put "could" in front of—at the beginning' (C). Focus group members particularly commented on the use of the word 'opinion', vehemently expressing the view that they were not interested in the witness's 'opinion', as if this indicated a subjective assessment where they were expecting and wanting an objective result that was not reliant on any form of human judgement: 'I think it's like, it's her interpretation' (C); 'once again, you're going on her opinion' (E1); 'she shouldn't be saying "my opinion", it should be "the statistics show"' (E1); and 'because it's only her opinion' (P1).

There was some acknowledgement that this idea that the DNA result was conclusive had come from 'the media' or 'television', especially programmes like *CSI* (mentioned at some point in most of the focus groups). Being presented with the LR statement and invited to say what they think it means precipitated (for some members in most focus groups) a dramatic (strongly voiced, with supportive agreements) loss of confidence in the evidence: 'if I was

25. 'She' refers to the expert witness presenting the evidence on the video-recorded scenario.

one of the jurors, it's going to give me a huge uncertainty from now on because I was totally, 100% agreeing with DNA as the last, last, or final straw [...] very conclusive [...] but now I'm kind of ...' (P2). This discourse was pivotal in structuring the talk, with 'not certain' being associated with 'not precise', 'not definite', 'not accurate', 'too broad', 'too general'. There was an expectation that the DNA result would be 'definite', like a fingerprint 'match' (which was assumed to be 100% certain, and 'scientific'). This reflects a reversal of the current thinking in the field: Saks and Faigman (2008) argue that fingerprints as 'forensic identification' is, in fact, a 'non-science' precisely because its 'certainty' can only be a construct of subjective conviction.

In their study, McQuiston-Surrett and Saks (2009) found mock jurors felt less uncertainty about the expert's findings regarding the identification of a hair when the expert gave a qualitative concluding 'opinion' that the defendant was the source. In our research, while there was indeed a reliance on the concluding 'opinion' as the statement that could be understood, and therefore the one on which there was more comfort to base a judgement, there was also considerable disquiet about the notion of an 'opinion' being the basis for this assessment. The loss of confidence in the testimony was reflected in additional comments such as: 'you guys are shattering my faith in DNA' (M2); 'it takes away your whole trust in scientific DNA profiling' (P1); and 'I don't trust it' (P1)—followed by 'It's not that I don't trust it, I just thought that DNA was a sure thing' (P1).

Confusion resulting from perceived ambiguity was compounded by confusion over the interpretation of the actual numbers presented in the LR statement. As Koehler wrote in 1996, '[n]umbers like one in one trillion boggle the mind' (p. 860), and elsewhere (2001) he reflects on the way jurors are not necessarily impressed by high numbers (low likelihood ratios). In our research, some were impressed with the sheer impact of a LR of one in a million, million, as we have shown above, yet at the same time the inability to make sense of the numbers was discussed in such a way by others that it compounded the confusion and reduced their confidence in their own ability to comprehend and form a judgement. Once their confidence in their comprehension was reduced, confidence in the process itself also declined. We give three examples of dialogue that reflect the kinds of interchange as focus group members grappled with the numbers. The first is from focus group P1[26]:

4(f): I don't think it's broad. Do you think one million million million is broad?
(f): Yeah. I don't think so either.
1(m): Million million million?
6(f): One million million or one million million million?
GM: It's 12 noughts.
(f): It's 12 noughts.
1(m): That's one billion. They can zoom it down to one person.
4(f): Yeah. One person unrelated to Mr. Jones...
1(f): Out of one million million.
4(f): unrelated to Mr. Jones

26. The code preceding the quotation refers to the number of the focus group member and in brackets the gender, for example 4(f) is focus group member number 4 who is female. Where no number appears, the group member was not able to be distinguished for that particular statement.

1(f): Well lets get his whole family in there.
4(f): I think so.

The second example is from focus group M2:

3(f): A thousand million times more likely. Blah blah blah.
1(m): How many, where is it? [trying to find it on the transcript]
3(f): 1,000 million times more likely that the DNA originated from Mr. Wilson.
1(m): You can never even count to that number.
3(f): Yeah, well that's the thing. Who's going to, you know, one, two, three, four….
1(f): You'd just switch off.

The third example is from focus group M2:

2(f): Does she mean six out of ten or six out of the twenty actual possible results?
5(f): It sounds like six out of twenty.
1(m): It's out of ten.
1(f): six out of ten sites.
JV: it was out of twenty.
2(f): Yeah, I thought it was the twenty. They didn't really, she didn't clarify that either.
3(f): What's twenty? Where did this come from?
2(f): You get two … .
1(f): …You get it twice…
2(f):…You get two results from each site that's tested, so you'll get twenty results, which means they only had three

of the ten sites originally, to test, if they only got six results.

1(f): What's that?

1(m): No, they had six of the ten sites.

2(f): No, six results of the twenty.

1(m): Oh.

3(f): See, we're confused, there you go. Don't know what she said. Six, ten, twenty, that's what I heard.

2(f): At each site they expect to get two results.

1(m): No, I thought it meant that they got six of the ten.

1(f): No, they got six results, not six sites.

2(f): So a DNA profile has twenty results. And this particular sample only got six, out of twenty.

1(m): Oh. I thought our DNA was whole. Straightaway anyway.

Overall, some were 'confused', 'lost', 'it's way too many numbers' (C), while others were unimpressed: 'The one thousand million, it's just blown it out of proportion basically' (P2); 'She's sort of embellished it by using that numerical term [a thousand million]' (M2); 'It's like she's exaggerating her statistics' (P2). This intersection of confusion (at times incomprehension), inability to generate an understanding, and getting lost is juxtaposed to the discourse of those (far fewer) for whom the uncertainty was to be expected because it is inherent in what it means to be 'scientific': 'but it's a scientific thing. They don't ever say "yes, you're definitely, it's positive"' (M2); and 'I do expect "it could be this, it could be that"' (C). It is noteworthy, however, that in the discussion these two perspectives at variance with one another travelled on separate paths: there was no productive interchange regarding their differences.

The professionals interviewed did not, in the main, express their own confusion but rather their concern about the confusion they believed jurors would experience when confronted with the LR statement. To jurors it 'loses meaning'; the 'figures are baffling'; as soon as 'partials and mixtures' are added it 'becomes very complicated'; 'sounds truly dreadful'; 'becomes astonishingly messy very, very quickly'; 'I don't think juries really understand'; 'it's really wordy, it's confusing'; and 'half the people out of the public wouldn't even know what it means'. There were a few exceptions where professionals referred to their own lack of comprehension. For example, one Crown Prosecutor said that, to him, the sheer magnitude of the numbers is meaningless; a medical doctor said that, if the LR was read to her she 'wouldn't take it in'; and one defence lawyer was intrigued to be presented with the opportunity to really wonder what it actually 'means'. This acknowledgement of a lack of understanding by a defence lawyer reinforces the overwhelming finding from a Norwegian study. Dahl's (2010) thematic analysis of the views of lawyers, particularly for the defence, with respect to the presentation of probabilistic DNA evidence in court shows a consistent concern across the group that could be summarized by two observations: the first relates to the absence of an understanding of the meaning of forensic DNA evidence; and the second, which follows, relates to the perception that it is infallible. Dahl analyses the 'black-boxing' involved in these perceptions.

Many of the concerns raised by the professional groups did, in fact, reflect various elements of the nature of the talk among the focus group participants. There was concern, for example, about whether some terms, such as 'random', would be misinterpreted rather than understood in the scientific sense intended (l.17). One Forensic Scientist emphasized that a shift from evidence presented as if it were 'fact' ('I saw the person cross the road') to evidence presented in terms of likelihoods or opinions might mean that 'the subtleties are lost on the jury' (l.1). A view was also expressed that jurors would be impressed by the 'big number', 'think "that must be him"' and 'I think they just go, "oh, it's one in a billion [...] the chances of it being anyone else are one in a billion", which is sort of true' (l.15); 'I think they would interpret that to mean yes, it's from this person' (l.3); 'I think they regard it as overwhelming evidence that's the person' (l.30); and 'probably all they want to hear is the very last sentence' (l.21).

The scientists interviewed focussed more attention on the detail of understanding, wondering how jurors might make sense of the LR statement. The Police personnel on the other hand, tended to voice a confidence that the jurors would listen to the last sentence and make up their minds on that basis. In the words of one Crown Prosecutor, 'but they get the message: it's him, we just can't say that' (l.27). Similarly, another Crown Prosecutor said, 'To read all that when you really just mean: this is Mr. X's' and 'they don't necessarily listen and understand all of that statistical stuff. But what they understand is, it's Mr. X's DNA' (l.28). An agonistic approach[27] to the presentation of evidence is fully consistent with the adversarial nature of the court proceedings; for example, the point was made by a Scene

27. The word 'agonistic' means a conflictual encounter with stakes attached and no predeterminable outcome; the outcome depends on the skill, daring and courage of the protagonists (its main usages are literary, and it comes from Greek, especially related to 'games', contests, etc.).

of Crime Officer that 'a good defence lawyer would add confusion to a jury' (I.5), implying that this would not be difficult. We note, however, that any such challenge to DNA evidence rarely happens in the New Zealand courts (JB)[28]. A Police detective's comment indicates a high degree of confidence in the impact of the opening and closing statements made by the Crown Solicitor in a case: 'they ignore all of that and they just simply say: "you're [the jury] going to hear evidence from the ESR which says that this guy is one hundred thousand million times more likely to be the offender than anyone else"' (I.6). If this does in fact happen in court, it represents direct use of a fallacious interpretation that, in some jurisdictions in the US, has been grounds for appeal and retrial (see Koehler, 1993; Murphy *et al.*, 2009). Another example of the confident use of the transposed probability by the prosecution is given by a Crown Prosecutor who insisted that the jury are impressed by the sequence of evidence: 'well, no, they [defendant] didn't admit they were there until *we got* a DNA hit saying that they were forty million times more likely to be that person—"oh yeah, well, OK, I was there"' (I.27). The patterns evident in these examples suggest how the mis-stating of statistically-presented LRs can be used in disconcertingly misleading ways.

The concerns regarding confusion invariably led to evaluative suggestions by focus group members and professionals on their preferred means of communicating DNA evidence. The following section briefly discusses the

---

28. JB refers to John Buckleton, Principal Scientist and statistician for ESR. John Buckleton was interviewed in his professional capacity as a key informant for this research.

points raised in this context, and the controversies they present.

'It Would Be Better If...'

Focus Groups

The points of tension generating some of the confusion in the sections discussed above crystallized in analysis of the juxtaposition of statements made about the preferred means of communication. On the one hand the only way the presentation of the evidence would be understood by lay members of the public as jurors was if it was considerably simplified and inevitably reduced: 'keeping it simple so that people will understand' (E1); 'just a little bit clearer' (M2); 'maybe just paraphrase' (M2); 'provide a little summary of what's going on so you don't get lost' (M2); and 'isn't there a different way of telling the story?' (P2). Implicit in these statements is a *desire* to understand, to follow, as if a narrative in simple words or images—'I think it should be pictorial' (M2)—would be possible and would lay out a pathway of conceptual logic that would be able to be followed and understood in common sense terms. Yet simultaneously, this simplification would almost certainly lead precisely to an insoluble dilemma: it would create a problematic and compromised position for jurors, and arguably for defendants and the CJS as a whole.

This would occur because simplification would lead to an absence and deferral of deeper understanding, creating a default position of relying on, or 'trusting', the experts. The simple wording of the evidence thus presented would invariably eliminate the grounds for uncertainty, effectively black-boxing the DNA typing evidence. This wish for simplification and the apparent certainty it reinforces appears in the following comments by focus group members: 'why doesn't she just come straight out and say "it's him"?' (M1); 'she could have just said "it was Mr. Wilson"' (M2); 'like, basically, "we took the DNA, and what it means is this: we're pretty certain it was him, but there's a something percent chance that it might not be", or something like that' (M2); and 'I think the answers could be a little bit more front end, front loaded, saying you know, "through the tests we have confirmed that the sample taken from the knife is from Mr. Wilson" (C). If we combine these observations with the discussion on the notion of 'opinion' above, it seems that the discourse of the focus group members both expresses a wish for a more simple statement of opinion at the same time as a wish for the opinion not to be an 'opinion' at all, but rather to represent the truth. In this sense, the desired simplification would make the evidence say what it does not say. Other more elaborate statements made by focus group members as they attempted to express (felt compelled to express) how the evidence could be more appropriately stated emphasize this point: 'one thing I would recommend is that they don't sit there and say "compared to the general public" […] whereas sitting there and saying "there is a 99% chance that this is more identifiable to Mr. Jones in comparison to the three suspects that we have", it's a bit more exact. And DNA is exact' (M1); 'I think it's important that they say "oh, by the way, 99.9999% of DNA cases have all been proven true", or 100%' (M1); and 'if they'd said "well, the sample that we got from Mr. Wilson has the likelihood of being one in six in the entire world" it would have been more… a bit better' (E1).

There were two notable statements that indicated a recognition of this dilemma and expressed a worry about the terms in which it is constructed, illustrated by M2: 'And so if they're turning up into court and using all this jargon, you tend to just sit back and think: oh, sweet as, they know what they're doing, I'll just wait for the ruling. And what I worry about is that in the midst of all this language and all the scientific testing is we lose our control, our ability to understand what's going on. So I'm probably getting more and more concerned as this goes on around, you know, it's like suddenly getting a bit too difficult and a bit too much in my head. I just have to have faith in people that don't know who I am…' (M2).

Professionals

There was a tendency across all professional groups to favour a minimalist approach to presenting evidence in court, in the interests of being 'understood' and not confusing the jury who 'need to take home a simple take-home message' (I.23): 'there has to be a way that, it has to be a lot more simpler to explain: it's either him or it's not' (I.5). Ambivalence does, however, transpire. Acknowledging the importance of the numbers, I.23 (a Police manager) states 'I mean there is very good reason why a scientist has to stand up in court and give all of this technical spiel because it has to be shown to be scientifically proven….' Assuming that the interpretation can indeed be 'proven', she seeks a way to both have the detail and yet give the jury the simplicity required to detour the possibility of what she would presumably regard as erroneous conclusions being drawn. Preferring an approach that is fully disclosing and against the trend to minimalism, a Forensic Scientist voiced the importance of the numbers being in the statement because they provide an explanation of 'how we get to the final figure and how we get to the final sentence' (I.21). A Forensic Scientist working as a crime scene specialist with ESR said 'I would argue that part of the forensic scientist's role is to select the information in such a way that you aid the jury's understanding' (I.17); for I.17 this aid, however, involves removing elements that might be unable to be understood and 'dumping a lot more in the appendix'. The result appears more able to be 'understood' because the apparent objectivity obscures the interpretive element intrinsic to the process, and the inherent uncertainty that is epistemologically necessary[29].

Oftentimes, when the DNA findings are not in dispute, the DNA analyst is not required to give evidence during the trial. On these occasions the analyst's statement of findings is read to the court by a clerk, which in New Zealand is referred to as 'having the evidence read'. Some professionals were concerned that the quantification should not be included when the statement is read out in the court. A Crown Prosecutor was adamant on this point: 'My view, when it comes to things of a scientific nature, is never, ever have that evidence read. Never, ever, *ever* have that evidence put in admission of fact' (I.27). I.27 is referring here to the evidence being inferential, and he wishes to use it rhetorically in the court, to draw out connections as a part of the means to make his case. The Scene of Crime Officer's point (mentioned above) that 'a good defence lawyer would add to the confusion of the

---

29. This interviewee acknowledges also that he does 'get accused by some of my colleagues in peer checking, of favouring minimalism in statements rather than going for more detail' (I.17).

jury' (I.5) also links the wish to simplify with the desire to secure the grounds for the prosecution: 'give them a figure, say it's one million, million to one that it's his ... it's got to be convincing, if they can get it down to that level'. Here I.5 is following the same problematic route as some focus group members' comments discussed above, in that simplification leads to erroneous statements being made, something I.5 possibly intuits when he follows this suggestion with the thought that then 'a defence lawyer will play on that'.

A Forensic Medical Practitioner took an extreme view that really goes to the heart of the problematic presented by this analysis. In his view, this kind of evidence should be out of the courtroom as such: 'that kind of thing, in my view, should not be in front of a jury' (I.18). He is taking a view that continues to be rehearsed in legal contexts: whether or not 'this kind of thing should be sorted out before the trial, between the panel of experts, and panel of lawyers and panel of judges, and whatever else. So that what is put in front of the jury is: this stuff came from this guy; this stuff did not come from this guy; or we just don't know.' We see here that there is no way around this dilemma reappearing: either jury members are requested to 'trust' the experts, or any version of the evidence that is 'simplified' (*ergo* 'certain') to be able to be 'understood' by lay members of the jury will in fact be erroneous. Furthermore, each approach will present the 'scientific' nature of the evidence as a rhetorical guarantor of truth in a manner that is precisely of the order of 'non-science'. Such a guarantor rhetorically persuades through belief in certainty in a context (the court) where science is introduced for the explicit purpose of presenting an alternative to the possibility of personal and group convictions entering into the judicial process. Thus, in each case, 'non-science' thinking can invade and construct 'common sense'. To treat 'science' as 'certain' would be taking us back to the episteme of the days (not so long ago in fact) when it was not unknown for the European courts to commit donkeys, swine, locusts and corpses to trial (Kadri, 2005).

FROM BELIEF-IN-CERTAINTY TO SCIENCE-AS-UNCERTAIN

We now summarize the main points emerging from the analysis of our empirical material and consider the implications for the use of the LR in court, reflecting as we do so on the notion of 'science' in the STS field.

The initial formulation of our objective to explore whether there is a gap, and if so, the nature of this gap, between the understandings of professionals and the lay public, has been partly supported, partly refined, and partly overturned. The view of the professional groups that the 'public' want certainty from forensic evidence is generally supported by the material analyzed; although there was a subordinate discourse to the contrary, it was not influential in disturbing the more complexly interwoven network of ideas consolidating the desire for, and expectation of, certainty[30]. The 'gap', however, does not open up between a group of CJS professionals who understand

---

30. Research by Podlas (2006) led her to conclude that there is no such thing as the 'CSI effect', which might harm the prosecution. Our research complicates this assertion, problematizing two of the definitional points that Podlas makes regarding the 'CSI effect': 1) It creates an unreasonable expectation on the part of jurors, making it more difficult for prosecutors to obtain convictions (there was an 'unreasonable expectation' for certainty among participants in our focus groups, some of which was actively attributed to *CSI* on television; it would be reductionist however, to attribute this expectation solely to this source), 2) *CSI* raises the stature of scientific evidence to virtual infallibility, thus making scientific evidence impenetrable (we question whether *CSI* programmes can be responsible for such a development and would rather see this portrayal as co-emergent with broader societal dynamics trending towards a simplification or even caricaturing of the meaning of knowledge).

the meaning of DNA evidence in accordance with the scientific and statistical theories generating this meaning, and a lay public who is confused, does not understand, and interprets erroneously. Rather, three distinctions have emerged:

1. *Science as uncertain.* For this group, there was an appreciation of the inevitability that there is an element of uncertainty in the presentation of scientifically assessed findings. There was an understanding that the probabilistic nature of the evidence does not represent an absolute truth. There was also an appreciation that the expert witness will offer an *opinion* in court (rather than a categorical statement of truth), and that this opinion will be restricted solely to the *strength of the evidence* as it points to the source of the crime scene DNA (rather than being an opinion regarding the identification of the accused or a statement of his or her guilt). This group was dominated by Forensic Scientists.

2. *Belief in science as certain, confirmed*—those for whom the evidence reflects certainty, even though it is presented in probabilistic terms, and that this is best represented in the 'opinion' statement, interpreted as an opinion regarding the identification of the accused (this group was dominated by the Police and personnel for the prosecution);

3. *Belief in the certainty of science undermined*—those for whom, contrary to their expectation, the evidence reflects uncertainty and as such this disempowers their ability to form a judgement or accept a

conditional opinion (mainly represented by the lay public focus group members).

This tripartite structure of relations to knowledge contains a series of oppositions that can be understood theoretically when indexed to the judicial context. The binary that structures the relation between Group 1 and Groups 2/3 could be described crudely as that of science-as-uncertain, grounded in a critical epistemology[31], versus a belief in a kind of mythical science that produces certainty. This binary opposes two notions: that science does not claim knowledge of what 'is', but works through a process of critique and on-going assessment acknowledging the role of judgement in relation to action on the one hand, and on the other hand that of a belief in the certainty of science that claims a conviction of what

31. According to Ulrich (1983), the term 'critical' has two distinct meanings in different paradigms within the philosophy of science: some, for example Popper (1959), use the word 'critical' to refer to an approach that opens propositions about reality to questioning, while others, for example Habermas (1979), use it to refer to an approach that enables dialogue on norms and values as well as propositions about reality. Although these two approaches are fundamentally opposed in some regards (Popper and other analytical philosophers of science exclude the exploration of norms and values from science), both paradigms nevertheless accept that science is inevitably uncertain. We use the term 'critical' in the phrase 'critical epistemology', however, to accord with a third approach that is not fully reflected in either the work of Popper or Habermas, and is best exemplified in the work of Latouche (1984). For Latouche, all scientific inquiry in the natural and social sciences progresses through a process of critique (through questioning discourses that can be shown to be false—for example, because they claim an absolute truth, or the evidence can be deconstructed), and with respect to the social domain, critique in the sense deriving from a Marxist notion of a critique of ideology.

'is', based on evidence that is 'revealed', unequivocal, and usually referenced to an external authority.

A second binary is an inverse relation between Group 2 and Group 3: both take the view that the knowledge generated through 'science' can be true and certain, whether this is framed as a means (tactically, rhetorically—the important thing about the truth claim is how it is stated and how its truth is pursued) or as an end (existentially—the important thing with the truth claim is that it is in fact true, that is, what it means), with Group 2 having this belief confirmed and Group 3 having it threatened and undermined. Group 1 discourse privileges the procedural; Group 2 discourse is agonistic; and Group 3 discourse is one thrown into disarray, unstable and open therefore to the agonistic rhetorics of both the prosecution and defence. The rhetorical use made of transposition by both professionals in Group 2 and focus group members in Group 3 served either to confirm the certainty that the defendant was the source of the crime scene sample, or to undermine this same certainty—both of which are problematic.

This depiction of the social, discursive relations at play in this research does not align with the assumed problematic of the demarcation of 'science' and 'common sense'. To endorse this latter binary is to assume a domain of science that is technical, objective, dealing with mathematical 'proofs' that have no correlates in the world of human judgement; it is also to assume a domain of human judgement called 'common sense' that is 'reasonable', based on life experience, ethically reflexive, and generally reliable as an index of social values. A sociological critique of this binary as it is enforced in the courts may work by deconstructing the assumptions related to science, showing that elements of common sense traverse the practice of science (as discussed earlier with reference to Lynch and McNally's analysis of the *Regina vs Adams* case). While endorsing this critique, we take it slightly further. The constructions of both 'science' and 'common sense' within this binary are problematic. We are in agreement that science does not function in the way this binary assumes. This prompts us, however, to suggest that how science does (should) work cannot be reduced to, or encompassed within, the order of 'common sense', if 'common sense' involves the uncritical acceptance of externally authorized 'truths'. On the contrary, science is neither relativist nor absolutist, and must importantly be demarcated from the construct of belief-in-certainty. In the sense described above, science works through critique, debate and controversy, and does not produce positive knowledge. In contrast, belief-in-certainty functions with convictions about truth, convictions that are (or can be) assumedly authorized by a mythical construct of 'science'. The very observation that various 'publics' might be losing their faith or confidence in science, or in Locke's (2001) words are 'increasingly ambivalent', is precisely the deviation that confirms this observation; in other words, this 'faith' can be dislocated from 'science' and directed elsewhere. We argue that there is an obscured and unacknowledged yet dynamic and structuring binary of science-as-critique that acknowledges uncertainty and belief-as-certainty that assumes absolute-truth, functioning in this domain of forensic science in court. While belief-in-certainty animates the agonistic practices of the adversarial system pivoting around an axis of prosecution *contra* defence, science is really not playing much of a role at all. Contrary to the view that 'science' *really* functions like 'common sense', it is our view that a 'common sense' aligned more actively with science-as-critique would enhance the potential for the science of DNA evidence to be a force for justice in the court system[32].

While we acknowledge that some authors are critical of the use of Bayesian LRs for the presentation of DNA evidence in court (see Lawless, 2010), our analysis inclines us to the view that any detouring away from this method at present is almost certainly going to reiterate the dilemma of undue simplification leading to erroneous statements that in turn obscures the subjective assessments integral to the statistical estimates it delivers. Some authors are proposing the use of 'Bayesian networks' as a means to making evidence understandable to lay jurors (see Foreman *et al.*, 2003). This may be a welcome development, although our research indicates a more fundamental understanding of the nature of science would need to precede such an intervention. As Koehler (2001) claims, the way the statistic is presented in court, its very wording, affects how jurors think about and evaluate the evidence. We asked the question earlier

32. It is important not to lose sight of defendants as those who are possibly the most affected personally by the politics of forensic knowledge. A recent study by Prainsack and Kitzberger (2009), based on interviews with prisoners in Austria, showed how a DNA 'match' is considered by this group to constitute a reality that they could in no way contest or avoid. The technologies were experienced as 'impenetrable and intimidating', creating a domain of evidence outside of their control. To the prosecution this might be a welcome prospect. From the point of view of our research, however, it reveals another source of potential manipulation of beliefs about truth and certainty that undermine the notion of justice.

whether the presence of fallacious reasoning matters. If it affects the assessment of evidence in this way, surely the answer must be 'yes, it does matter'[33]. We take note of the advice proffered by Lambert and Evett twelve years ago (1998) that the wording of the final 'opinion' statement (an opinion perfectly legitimate, indeed necessary, to make) can be misleading. They are cautious about the way forward on this point, but suggest an additional statement could be made to the effect that 'the above opinion relates to the strength of the DNA evidence and does not represent an opinion on the origin of the [crime scene sample]' (p. 270). We endorse this addition, and in line with their recommendations we consider that this final inference statement in the LR statement presented by ESR in New Zealand could usefully be changed to claim that the opinion is based on a 'scientific assessment' rather than that the evidence provides strong (or extremely strong) 'scientific support'. Presumably, the same sound scientific methods have been used when the expert witness says that the evidence is strong as when he or she says it is weak. The statement should also be adjusted further to mean that the two hypotheses are clearly related to the likelihood of obtaining the profiling results[34]. We

33. We note Buckleton, Triggs and Walsh's (2005) analysis of the mathematical consequences of the prosecutor's fallacy, where they point out that the consequences vary depending on the prior odds.

34. The current version reads: 'I have compared the likelihood of two possible alternatives: either the DNA present in this sample originated from Mr. X, or the DNA originated from another male, unrelated to Mr. X, selected at random…' If this were amended to add 'the DNA crime scene sample profiling results would have been obtained if' after 'either' and 'or' (so it would read: 'I have compared the likelihood of two possible alternatives: either the DNA crime scene sample profiling results would have been obtained if the DNA present in this sample originated from Mr. X, or the DNA crime scene sample profiling results would have been

consider that such emendation would be supported by
the Amici Curiae group (Murphy, 2009) who scrutinize and
criticize court proceedings in regard to their deliberation
on DNA evidence. The direction we outline is entirely
opposed to the removal of the word 'opinion' from the
final statement, and considers the suggestion of removing
the presentation of DNA evidence from the courts to be a
'black-boxing' gesture that reinforces closure in the sense
meant by Halfon (1998), discussed earlier. On the basis of
our analysis we also consider it advisable for the expert
witness to present to jury members in court a generic
written statement that explains the interpretation of
forensic DNA evidence using the LR.

Jurors are required to make a judgement regarding guilt
or innocence. This research suggests that jurors' desire
for expert certainty may function to displace the onus
of truth in the face of inevitable uncertainty onto an
external authority, thus relieving jurors from taking this
responsibility on themselves. As long as jurors, as members
of the lay public, largely assess evidence on grounds
pertaining to belief in the possibility of certainty through
science ('belief' with its etymological connotations of
servility), the jury's judgement will be an epiphenomenon
of the argumentative drama in the courtroom that
plays on and with these beliefs. A more sophisticated
understanding of science-as-critique[35] would provide

obtained if the DNA originated from another male, unrelated to Mr.
X, selected at random...'), this would reduce the tendency to abstract
the hypotheses from their accurate context, and also give greater
leverage to the defence to insist on an accurate and non-transpositional
interpretation.

35. We insist that our view here departs from the 'deficit model', which is
rightly critiqued by sociologists (see Locke, 2002). The deficit model takes

the grounds for the jury to refuse any usurpation of their prerogative to make their own independent judgement, and to cease any tendency to act like a firing squad[36]. Nevertheless, we recognize that the current reality is a long way from what we argue is the ideal of juries coming into the courtroom with an existing understanding of science-as-critique. Indeed, there is a question mark around whether it will ever be possible to realize this ideal in practice, given that most members of the public do not, and will not want to, study epistemology. We therefore argue that the emphasis needs to be on improving the current situation through small steps, and creating the conditions for juries to take responsibility for what they understand to be a matter of judgement. Continued work to enhance the presentation of evidence (noting our suggestions above) is required, as well as innovations in the university education of professionals starting careers in the CJS. Developing a thorough appreciation amongst lawyers of how the acknowledgement of uncertainty is central to the operation of science is going to be critical in helping them better communicate the reasons for probabilistic statements to juries who might be expecting pronouncements of certainty. A good knowledge of the

---

the myth of science as productive of certainty and constructs the 'public' as 'ignorant' or lacking the scientific literacy necessary to understand this production. Rather than reinforce beliefs about 'science' through an uncritical education, our preference to a relativist alternative is to broaden and deepen more general societal understanding of science as a critical, endlessly questioning, entirely situated endeavour.

36. In her history of the trial, Sadakat Kadri (2005) makes the point that 'the primary role of the people in the jury box has never been to establish ultimate truths. It has been to carry the can […] small enough for collective liability but large enough to lose individual responsibility, they are pressed to serve as a firing squad' (p. 342).

main statistical procedures used in the production of evidence is also essential, so they can explicitly address any tendency to transpose and misinterpret statistical statements. Finally, as part of their basic education, it is important for lawyers to understand how 'black-boxing' prevents critique of the processes by which evidence is produced (other than in the highest profile cases). These days, given the increasing use of forensic DNA evidence in court, it is arguably the duty of lawyers to keep abreast of the major debates on the construction of evidence that are taking place in the scientific community; in some cases it might be in the interests of justice to bring these debates into the courtroom rather than allow them to be resolved (black-boxed) behind the closed doors of scientific committees. The practical implications indicated by this research are reinforced by Dahl's (2010) interpretations of the findings from her interviews with lawyers in the Norwegian context.

Our final point is that a social study of science risks impotence and relativism if it reduces its scope to a critique of the positive rendition of science (that is, to a mythical[37] science *as* belief-in-certainty, science that can be 'black-boxed') and does not take up the project of what a critical, subjective and reflexive, messy science in both its natural and social incarnations, motivated always from a standpoint of action, might be. Such a view of the way science works enables knowledge claims to be evaluated in a way that does not succumb to belief in epistemic certainty and truth.

---

37. We allude here to Girard's (1979, c. 1972) notion of myth as a story that effects an obscuring of a foundational murder, which in turn covers over an unbearable lack, or absence.

REFERENCES

Braun, V. and Clarke, V. (2006). "Using thematic analysis in psychology," *Qualitative Research in Psychology*, ISSN 1478-0887, 3(2): 77-101.

Buckleton, J.S., Triggs, C.M. and Walsh, S.J. (2005). *Forensic DNA Evidence Interpretation*, ISBN 9780849330179.

Cole, S.A. (2009). "Forensics without uniqueness, conclusions without individualization: The new epistemology of forensic identification," *Law, Probability and Risk*, ISSN 1470-840X, 8: 233-55.

Cole, S.A. (2004). "Fingerprint identification and the criminal justice system," in D. Lazer (ed.), *DNA and the Criminal Justice System: The Technology of Justice*, ISBN 9780262621861, pp 63-91.

Dahl, J.Y. (2010). "DNA the Nor-way: Black-boxing the evidence and monopolizing the key," in R. Hindmarsh and B. Prainsack (eds.), *Genetic Suspects: Global Governance of Forensic DNA Profiling and Databasing*, ISBN 9780521519434, pp 197-217.

Dahl, J.Y. (2009a). "Another side of the Story. Lawyers views on DNA as evidence," in K.F. Aas, H.O. Gundhus and H.M. Lomell (eds.), *Technologies of InSecurity. The Surveillance of Everyday Life*, ISBN 9780203891582, pp 219-237.

Dahl, J.Y. and Sætnan, A.R. (2009b). "'It all happened so slowly': On controlling function creep in forensic DNA databases," *International Journal of Law, Crime and Justice*, ISSN 1756-0616, 37 (3): 83-103.

Derksen, L. (2010). "Micro/macro translations: The production of new social structures in the case of DNA profiling," *Sociological Inquiry*, ISSN 0038-0245, 80(2): 214-40.

Derksen, L. (2000). "Towards a sociology of measurement: The meaning of measurement error in the case of DNA profiling," *Social Studies of Science*, ISSN 0306-3127, 30(6): 803-45.

Foreman, L.A., Champod, C., Evett, I.W., Lambert, J.A. and Pope, S. (2003). "Interpreting DNA evidence: A review," *International Statistical Review*, ISSN 0306-7734, 71(3): 473-95.

Giddens, A. (1997). *The Consequences of Modernity*, ISBN 9780804718912.

Gilbert, N. (2010). "DNA's identity crisis," *Nature,* ISSN 0028-0836, 464(18 March): 347-8.

Girard, R. (1979, c 1972). *Violence and the Sacred,* ISBN 9780801822186.

Habermas, J. (1979, c 1976). *Communication and the Evolution of Society*, ISBN 9780807015131.

Halfon, S.l (1998). "Collecting, testing and convincing: forensic DNA experts in the courts," *Social Studies of Science,* ISSN 0306-3127, 28(5-6): 801-28.

Holmgren, J. (2005). "DNA evidence and jury comprehension," *Canadian Society of Forensic Science Journal,* ISSN 0008-5030, 38(3): 123-41.

Holstein, J. and J.E. Gubrium (eds.) (2003). *Inside Interviewing: New Lenses, New Concerns*, ISBN 9780761928515.

Ivkovic, S.K. and Valerie P.H. (2003). "Jurors' evaluations of expert testimony: Judging the messenger and the message," *Law and Social Inquiry*, ISSN 0897-6546, 28(2): 441-82.

Jasanoff, S (1998). "The eye of everyman: Witnessing DNA in the Simpson trial," *Social Studies of Science*, ISSN 0306-3127, 28(5-6): 713-40.

Jasanoff, S. (1995). *Science at the Bar: Law, Science and Technology in America*, ISBN 9780674793026.

Jasanoff, S. (1990). *The Fifth Branch: Science Advisors as Policymakers*, ISBN 9780674300620.

Kadri, S. (2005). *The Trial: A History, from Socrates to O.J. Simpson*, ISBN 9780375505508.

Kaye, D.H. (2009) "Identification, individualization and uniqueness: What's the difference?" *Law, Probability and Risk* 8(6): 85-94. Online ISSN 1470-840X. Print ISSN 1470-8396.

Kaye, D.H., Hans, V.P., Dann, B.M., Farley, E. and Albertson, S. (2007). "Statistics in the jury box: How jurors respond to mitochondrial DNA match probabilities," *Journal of Empirical Legal Studies*, ISSN 1740-1453, 4(4): 797-834.

Kaye, D.H. and Koehler, J.J. (2003). "The misquantification of probative value," *Law and Human Behavior*, ISSN 0147-7307, 27(6): 645-59.

Koehler, J.J. (2001). "The psychology of numbers in the courtroom: How to make DNA match statistics seem impressive or insufficient," *Southern California Law Review*, ISSN 0038-3910, 74: 1275-1305.

Koehler, J.J. (1996). "On conveying the probative value of DNA evidence: Frequencies, likelihood ratios, and error rates," *Colorado Law Review*, ISSN 0041-9516, 67: 859-86.

Koehler, J.J., Chia, A. and Lindsey, S. (1995). "The random match probability in DNA evidence: Irrelevant and prejudicial?" *Jurimetrics Journal*, ISSN 0897-1277, 35: 201-19.

Lambert, J.A. and Evett, I.W. (1998). "The impact of recent judgements on the presentation of DNA evidence," *Science and Justice*, ISSN 1355-0306, 38(4): 260-70.

Latouche, S. (1984). *Le Procès de la Science Sociale* (Putting the Process of Social Science on Trial), ISBN 9782715710863.

Lawless, C. (2010). "Managing epistemic risk in forensic science: Sociological aspects and issues," *Sociology Compass*, ISSN 1751-9020, 4(6): 381-92.

Levesque-Lopman, L. (2000). "Listen, and you will hear: Reflections on interviewing from a feminist phenomenological perspective,"in L. Fisher and L. Embree (eds.), *Feminist Phenomenology*, ISBN 9789048155637.

Locke, S. (2002). "The public understanding of science: A rhetorical invention," *Science, Technology & Human Values*, ISSN 0162-2439, 27(1): 87-111.

Locke, S. (2001). "Sociology and the public understanding of science: From rationalization to rhetoric," *British Journal of Sociology*, ISSN 0007-1315, 52(1): 1-18.

Luhmann, N. (1989, c.1986). *Ecological Communication*, John Bednarz (trans.), ISBN 9780226496511.

Lynch, M. and Cole, S.A. (2005). "Science and technology studies on trial: Dilemmas of expertise," *Social Studies of Science*, ISSN 0306-3127, 35(2): 269-311.

Lynch, M. and McNally, R. (2003). "'Science','common sense', and DNA evidence: A legal controversy about the public understanding of science,"' *Public Understanding of Science*, ISSN 1361-6609, 12: 83-103.

Lynch, M. and McNally, R. (1999). "Science, common sense and common law: Courtroom inquiries and the public understanding of science," *Social Epistemology*, ISSN 0269-1728, 13(2): 183-196.

Lynch, M. (1998). "The discursive production of uncertainty: The O.J. Simpson "dream team" and the sociology of knowledge machine," *Social Studies of Science*, ISSN 0306-3127, 28(5-6): 829-68.

Macdonald, F. (2005). "Meanings of evidence: A scoping study," Summer Studentship Report, University of Canterbury (confidential document held by ESR).

McQuiston, D. and Saks, M.J. (2009). "The testimony of forensic identification science: What expert witnesses say and what factfinders hear," *Law and Human Behavior*, ISSN 1573-661X, 33: 436-53.

Murphy, E., Thompson, W.C., *et al.*, (2009). Brief of 20 Scholars of Forensic Evidence as Amici Curiae Supporting Respondents: *McDaniel v Brown*, No 08-559 in the Supreme Court (July 24, 2009).

Podlas, K. (2006). "The CSI Effect: Exposing the media myth," *Fordham Intellectual Property, Media and Entertainment Law Journal*, ISSN 1079-9699, 16: 429-64.

Popper, K.R. (1959, c 1936). *The Logic of Scientific Discovery*, ISBN 9780415278447.

Prainsack, B. and Kitzberger, M. (2009). "DNA behind bars: Other ways of knowing forensic DNA technologies," *Social Studies of Science*, ISSN 0306-3127, 39(1): 51-79.

Saks, M.J. and Koehler, J.J. (2008). "The individualization fallacy in forensic science evidence," *Vanderbilt Law Review*, ISSN 0042-2533, 61(1): 199-217.

Saks, M.J. and Faigman, D.L. (2008). "Failed forensics: How forensic science lost its way and how it might yet find it," *Annual Review of Law and Social Science*, ISSN 1550-3585, 4: 149-71.

Saks, M.J. (1997-8). "Merlin and Solomon: Lessons from the law's formative encounters with forensic identification science," *Hastings Law Journal*, ISSN 0017-8322, 49: 1069-41.

Thompson, W.C. and Schumann, E.L. (1987). "Interpretation of statistical evidence in criminal trials: The prosecutor's fallacy and the defense attorney's fallacy," *Law and Human Behavior*, ISSN 0147-7307, 11(3): 167-87.

Ulrich, W. (1983). *Critical Heuristics of Social Planning: A New Approach to Practical Philosophy*, ISBN 9780471953456.

Webster, D.B. (1999). *Neuroscience of Communication*, ISBN 9781565939851.

Yearly, S. (2004). *Making Sense of Science,* ISBN 9780803986923.

APPENDIX

Participant	Role
Interview 01	Forensic Scientist—Crime Scene Specialist
Interview 03	Forensic Scientist—Crime Scene Specialist
Interview 04	Police—SOCO
Interview 05	Police—SOCO
Interview 06	Police—Detective
Interview 07	Police—Detective
Interview 08	Forensic Medical Practitioner
Interview 09	Forensic Scientist—DNA
Interview 10	Police—SOCO
Interview 11	Police—Detective
Interview 12	Police—Detective
Interview 13	Police—Detective
Interview 14	Forensic Medical Practitioner
Interview 15	Forensic Scientist—DNA
Interview 16	Forensic Scientist—Crime Scene Specialist
Interview 17	Forensic Scientist—Crime Scene Specialist
Interview 18	Forensic Medical Practitioner
Interview 20	Forensic Scientist—DNA
Interview 21	Forensic Scientist—Crime Scene Specialist
Interview 22	Forensic Medical Practitioner
Interview 23	Police Management
Interview 24	Defence Lawyer
Interview 25	Police—Detective
Interview 26	Scientist—Defence Analyst
Interview 27	Prosecutor
Interview 28	Prosecutor
Interview 29	Prosecutor
Interview 30	Defence lawyer

Table 1 *Interview Participants' Interview Number And Role.*

Role	Total (n)	Gender		Experience		
		Males (n)	Females (n)	Less than 5 years	Between 5 & 20 years	More than 20 years
Defence Lawyers	2	2				2
Forensic Medical Practitioners	4	2	2		2	2
Forensic Scientists—Crime Scene Specialists	6	1	5	1	3	2
Forensic Scientists—DNA	3		3		2	1
Forensic Scientists—Defence analysts	1	1				1
Police—Detectives	6	6		1		5
Police—SOCO	3	1	2	1	1	1
Prosecutors	3	2	1	1	1	1

Table 2 Interview Participants' Role, Gender And Length Of Experience

Group	Number in Group	Females	Males	Location	Duration (minutes)	Facilitator	Notes
Māori M1	10	7	3	Ōrākei Marae	94	AAD & JV	Extended Family Group, wide age range.
Pacific Island P1	8	7	1	Unitec Institute of Technology	126	GM & JV	Staff from the Centre of Pacific Development And Support. Participants are from various Pacific Islands.
European NZ E1	8	5	3	Private Residence	96	JV	A group of people known to each other, wide age range. Most are Caucasian, 1 Māori, 1 Pacific Island.
European NZ E2	9	7	3	Avondale Returned Services Association	78	AAD & JV	A group of senior citizens
Māori M2	6	5	1	ESR	118	AAD & JV	A group of people known to each other, organized by a friend of AAD.
Chinese C	8	5	3	ESR	102	VG & JV	A group of immigrants, wide age range.
Pacific Island P2	5	3	2	Avondale Union Church	112	GM & JV	A Pacific Island group connected through a local church, wide age range.

Table 3 *Focus Groups Summary*